Con

M000284469

Why Scientific Thinking?

The Problem:

Everyone thinks; it is our nature to do so. But much of our thinking, left to itself, is biased, distorted, partial, uninformed, or down-right prejudiced. Yet the quality of our life and that of what we produce, make, or build depends precisely on the quality of our thought. Shoddy thinking is costly, both in money and in quality of life. Excellence in thought, however, must be systematically cultivated.

A Definition:

Scientific thinking is that mode of thinking — about any scientific subject, content, or problem — in which the thinker improves the quality of his or her thinking by skillfully taking charge of the structures inherent in thinking and imposing intellectual standards upon them.

The Result:

A well cultivated scientific thinker:

- raises vital scientific questions and problems, formulating them clearly and precisely;
- gathers and assesses relevant scientific data and information, using abstract ideas to interpret them effectively;
- comes to well-reasoned scientific conclusions and solutions, testing them against relevant criteria and standards;
- thinks openmindedly within convergent systems of scientific thought, recognizing and assessing scientific assumptions, implications, and practical consequences; and
- communicates effectively with others in proposing solutions to complex scientific problems.

Scientific thinking is, in short, self-directed, self-disciplined, self-monitored, and self-corrective. It presupposes assent to rigorous standards of excellence and mindful command of their use. It entails effective communication and problem solving abilities as well as a commitment to developing scientific skills, abilities, and dispositions.

The Elements of Scientific Thought

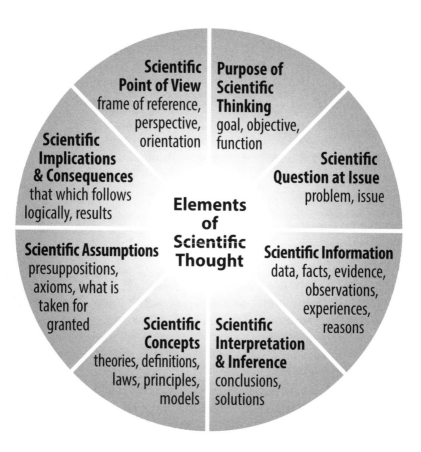

Used With Sensitivity to Universal Intellectual Standards

Clarity → Accuracy → Depth → Breadth → Significance

Precision

Relevance

↓

Fairness

Questions Using the
Elements of Scientific Thought
(in a scientific paper)

Scientific Purpose	What am I trying to accomplish? What is my central aim? My purpose?
Scientific Questions	What question am I raising? What question am I addressing? Am I considering the complexities in the question?
Scientific Information	What data am I using in coming to that conclusion? What information do I need to settle the question? What evidence is relevant to the question?
Scientific Inferences/ Conclusions	How did I reach this conclusion? Is there another way to interpret the information?
Scientific Concepts	What is the main concept, theory, or principle here? Can I explain the relevant theory?
Assumptions	What am I taking for granted? What assumption has led me to that conclusion?
Implications/ Consequences	What are the implications of the data I have collected? What are the implications of my inferences?
Points of View	From what point of view am I looking at this issue? Is there another point of view I should consider?

A Checklist for Scientific Reasoning

1) All scientific reasoning has a PURPOSE.

- Take time to state your purpose clearly.
- Distinguish your purpose from related purposes.
- Check periodically to be sure you are still on target.
- Choose significant and realistic scientific purposes.

2) All reasoning is an attempt to figure something out, to settle some scientific QUESTION, to solve some scientific PROBLEM.

- State the question at issue clearly and precisely.
- Express the question in several ways to clarify its meaning and scope.
- Break the question into sub-questions.
- Distinguish questions that have definitive answers from those that are a matter of opinion and from those that require consideration of multiple viewpoints.

3) All scientific reasoning is based on ASSUMPTIONS.

- Clearly identify your assumptions and determine whether they are justifiable.
- Consider how your assumptions are shaping your point of view.

4) All scientific reasoning is done from some POINT OF VIEW.

- Identify your point of view.
- Seek other points of view and identify their strengths as well as weaknesses.
- Strive to be fairminded in evaluating all scientific points of view.

5) All scientific reasoning is based on DATA, INFORMATION and EVIDENCE.

- Restrict your claims to those supported by the available data.
- Search for information that opposes your position as well as information that supports it.
- Make sure that all information used is clear, accurate and relevant to the question at issue.
- Make sure you have gathered sufficient information.

6) All scientific reasoning is expressed through, and shaped by, scientific CONCEPTS and IDEAS.

- Identify key scientific concepts and explain them clearly.
- Consider alternative concepts or alternative definitions of concepts.
- Make sure you use concepts with precision.

7) All scientific reasoning entails INFERENCES or INTERPRETATIONS by which we draw scientific CONCLUSIONS and give meaning to scientific data.

- Infer only what the evidence implies.
- Check inferences for their consistency with each other.
- Identify assumptions underlying your inferences.

8) All scientific reasoning leads somewhere or has IMPLICATIONS and CONSEQUENCES.

- Trace the implications and consequences that follow from your reasoning.
- Search for negative as well as positive implications.
- Consider all possible consequences.

Scientific Thinking Seeks to Quantify, Explain, and Predict Relationships in Nature

The true scientific investigator never jumps at conclusions, never takes anything for granted, never considers his judgment better than his information, and never substitutes opinions or long established belief for fact. No matter how plausible a given statement may be or how logical a proposed explanation of it may seem, it must be treated merely as a supposition until it has been proved true by searching tests. Moreover, these tests must be of such kind that other scientists can repeat them, and of such nature that others repeating them will inevitably come to the same conclusions. Only in this maner can a body of dependable scientific knowledge be built up.

Lincoln Library of Essential Information, 1940

Scientific thinking is based on a belief in the intelligibility of nature, that is, upon the belief that the same cause operating under the same conditions, will result in the same effects at any time. As a result of this belief, scientists pursue the following goals.

1. They Observe. (**What conditions seem to affect the phenomena we are observing?**) In order to determine the causal relations of physical occurrences or phenomena, scientists seek to identify factors that affect what they are studying.

2. They Design Experiments. (**When we isolate potential causal factors, which seem to most directly cause the phenomena, and which do not?**) In scientific experiments, the experimenter sets up the experiment so as to maintain control over all likely causal factors being examined. Experimenters then isolate each variable and observe its effect on the phenomena being studied to determine which factors are essential to the causal effect.

3. They Strive for Exact Measurement. (**What are the precise quantitative relationships between essential factors and their effects?**) Scientists seek to determine the exact quantitative relationships between essential factors and resulting effects.

4. They Seek to Formulate Physical Laws. (**Can we state the precise quantitative relationship in the form of a law?**) The quantitative cause-effect relationship, with its limitations clearly specified, is known as a physical law. For example, it is found that for a constant mass of gas, at a constant

temperature, the volume is inversely related to the pressure applied to it; in other words, the greater the pressure the less the volume — the greater the volume the less the pressure. This relationship is constant for most gases within a moderate range of pressure. This relationship is known as *Boyle's Law*. It is a physical **law** because it *defines* a cause-effect relationship, but it does not *explain* the relationship.

5. **They Study Related or Similar Phenomena.** (When we examine many related or similar phenomena, can we make a generalization that covers them all?) A study of many related or similar phenomena is typically carried out to determine whether a generalization or hypothesis can be formulated that accounts for, or explains, them all.

6. **They Formulate General Hypotheses or Physical Theories.** A theoretical generalization is formulated (if one is found to be plausible). For example, the *kinetic theory* of gas was formulated to explain what is documented in *Boyle's Law*. According to this theory, gases are aggregates of discrete molecules that incessantly fly about and collide with themselves and the wall of the container that holds them. The smaller the space they are forced to occupy, the greater the number of collisions against the surfaces of the space.

7. **They Seek to Test, Modify, and Refine Hypotheses.** If a generalization is formulated, scientists test, modify, and refine it through comprehensive study and experimentation, extending it to all known phenomena to which it may have any relation, restricting its use where necessary, or broadening its use in suggesting and predicting new phenomena.

8. **When Possible, Scientists Seek to Establish General Physical Laws as well as Comprehensive Physical Theories.** General physical laws and comprehensive physical theories are broadly applicable in predicting and explaining the physical world. The *Law of Gravitation*, for example, is a general physical law. It states that every portion of matter attracts every other portion with a force directly proportional to the product of the two masses,

and inversely proportional to the square of the distance between the two. Darwin's *Theory of Evolution* according to natural selection is a comprehensive physical theory. It holds that all species of plants and animals develop from earlier forms by hereditary transmission of slight variations in successive generations and that natural selection determines which forms will survive.

Universal Intellectual Standards Essential to Sound Scientific Thinking

Universal intellectual standards are standards which must be applied to thinking whenever one is interested in checking the quality of reasoning about a problem, issue, or situation. To think scientifically entails having command of these standards. While there are a number of universal standards, we focus here on some of the most significant:

Clarity:

Could you elaborate further on that point? Could you express that point in another way? Could you give me an illustration? Could you give me an example?

Clarity is a gateway standard. If a statement is unclear, we cannot determine whether it is accurate or relevant. In fact, we cannot tell anything about it because we don't yet know what it is saying.

Accuracy:

Is that really true? How could we check that? How could we find out if that is true?

A statement can be clear but not accurate, as in "Most creatures with a spine are over 300 pounds in weight."

Precision:

Could you give me more details? Could you be more specific?

A statement can be both clear and accurate, but not precise, as in "The solution in the beaker is hot." (We don't know how hot it is.)

Relevance:

How is that connected to the question? How does that bear on the issue? A statement can be clear, accurate, and precise, but not relevant to the question at issue.

If a person who believed in astrology defended his/her view by saying "Many intelligent people believe in astrology," their defense would be clear, accurate, and sufficiently precise, but irrelevant. (For example, at one time many intelligent people believed the earth was flat.)

Depth:

How does your answer address the complexities in the question? How are you taking into account the problems in the question? Are you dealing with the most significant factors?

A statement can be clear, accurate, precise, and relevant, but superficial (that is, lack depth). For example, the statement "Just Say No" which is often used to discourage children and teens from using drugs, is clear, accurate, precise, and relevant. Nevertheless, it lacks depth because it treats an extremely complex issue, the pervasive problem of drug use among young people, superficially. It fails to deal with the complexities of the issue.

Breadth:

Do we need to consider another point of view? Is there another way to look at this question? What would this look like from the point of view of a conflicting theory, hypothesis or conceptual scheme?

A line of reasoning may be clear, accurate, precise, relevant, and deep, but lack breadth (as in an argument from either of two conflicting theories, both consistent with available evidence).

Logic:

Does this really make sense? Does that follow from what you said? How does that follow? Before you implied this and now you are saying that? I don't see how both can be true.

When we think, we bring a variety of thoughts together into some order. When the combination of thoughts are mutually supporting and make sense in combination, the thinking is "logical." When the combination is not mutually supporting, is contradictory in some sense, or does not "make sense," the combination is "not logical." In scientific thinking, new conceptual schemes become working hypotheses when we deduce from them logical consequences which can be tested by experiment. If many of such consequences are shown to 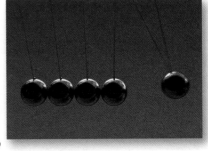 be true, the theory (hypothesis) which implied them may itself be accepted as true.

Intellectual Standards
in Scientific Thinking

Clarity

Understandable, the meaning can be grasped
Could you elaborate further on our hypothesis (or idea)?
Could you give me a more detailed explanation of the
phenomenon you have in mind?

Accuracy

Free from errors or distortions, true
How could we check on those data?
How could we verify or test that theory?

Precision

Exact to the necessary level of detail
Could you be more specific? Could you give me more
details on the phenomenon? Could you be more exact
as to how the mechanism takes place?

Relevance

Relating to the matter at hand
How do those data relate to the problem? How do they
bear on the question?

Depth

**Containing complexities and multiple
interrelationships**
What factors make this a difficult scientific problem?
What are some of the complexities we must consider?

Breadth

Encompassing multiple viewpoints
Do we need to look at this from another perspective?
Do we need to consider another point of view? Do we
need to look at this in other ways?

Logic

The parts make sense together, no contradictions
Are all the data consistent with each other? Are these
two theories consistent? Is that implied by the data we
have?

Significance

Focusing on the important, not trivial
Is this the central idea to focus on? Which set of data is
most important?

Fairness

Justifiable, not self-serving or one-sided
Do I have any vested interest in this issue which keeps
me from looking at it objectively? Am I misrepresenting
a view with which I disagree?

The Figuring Mind

Thinking Scientifically

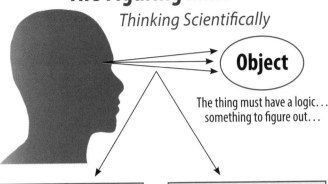

Object

The thing must have a logic...
something to figure out...

There is a logic to figuring something out scientifically, to constructing a system of meanings which makes sense of something

There are **intellectual standards** scientists use to assess whether the logic in their mind mirrors the logic of the thing to be understood

The Elements of Thought reveal the logic:

1	An object to be figured out	→ some data or information, some experience of it (the **Empirical Dimension**)
2	Some reason for wanting to figure it out	→ our **Purpose** or **Goal**
3	Some question or problem we want solved	→ our **Question at Issue**
4	Some initial sense of the object (whatever we take for granted)	→ our **Assumptions**
5	Some ideas by which we are making sense of the object	→ the **Conceptual Dimension**
6	Some drawing of conclusions about the object	→ our **Inferences** or interpretations
7	What follows from our interpretation of the object	→ the **Implications** and **Consequences**
8	Some viewpoint from which we conceptualize the object	→ our **Point of View** or **Frame of Reference**

Intellectual Standards include:

Clarity

Precision

Relevance

Accuracy

Depth

Breadth

Logic

Fairness

How to Analyze the Logic of a Scientific Article, Essay, or Chapter

One important way to understand a scientific essay, article or chapter is through the analysis of the structure of the author's reasoning. Once you have done this, you can evaluate the author's reasoning using intellectual standards (see page 20). Here is a template to follow:

1) The main **purpose** of this scientific article is _____.
   (Here you are trying to state, as accurately as possible, the author's intent in writing the article. What was the author trying to accomplish?)

2) The key **question** that the author is addressing is
   _____. (Your goal is to figure out the key question that was in the mind of the author when he/she wrote the article. What was the key question addressed in the article?)

3) The most important **information** in this scientific article is
   _____. (You want to identify the key information the author used, or presupposed, in the article to support his/her main arguments. Here you are looking for facts, experiences, and/or data the author is using to support his/her conclusions.)

4) The main **inferences** in this scientific article are _____
   _____.
   (You want to identify the most important conclusions the author comes to and presents in the article).

5) The key **concept**(s) we need to understand in this scientific article is (are)
   _____By these concepts the author means ____
   _____. (To identify these ideas, ask yourself: What are the most important ideas that you would have to know to understand the author's line of reasoning? Then briefly elaborate what the author means by these ideas.)

6) The main **assumption**(s) underlying the author's thinking is (are) _____ (Ask yourself: What is the author taking for granted [that might be questioned]? The assumptions are generalizations that the author does not think he/she has to defend in the context of writing the article, and they are usually unstated. This is where the author's thinking logically begins.)

7a) If we accept this line of reasoning (completely or partially), the **implications** are _____. (What consequences are likely to follow if people take the author's line of reasoning seriously? Here you are to pursue the logical implications of the author's position. You should include implications that the author states, and also those that the author does not state.)

7b) If we fail to accept this line of reasoning, the **implications** are _____. (What consequences are likely to follow if people ignore the author's reasoning?)

8) The main **point(s) of view** presented in this scientific article is (are) _____. (The main question you are trying to answer here is: What is the author looking at, and how is he/she seeing it? For example, in this thinker's guide, we are looking at scientific thinking and seeing it "as requiring intellectual discipline and the development of intellectual skills.")

If you truly understand these structures as they interrelate in a scientific article, essay or chapter, you should be able to empathically role-play the thinking of the author. Remember, these are the eight basic structures that define all reasoning. They are the essential elements of scientific thought.

Analyzing the Logic of a Science Textbook

Just as you can understand a scientific essay, article, or chapter by analyzing the parts of the author's reasoning, so can you figure out the system of ideas within a scientific textbook by focusing on the parts of the author's reasoning within the textbook. To understand the parts of the textbook author's reasoning, use this template:

The Logic of a Science Textbook

1) The main **purpose** of this textbook is _____.

2) The key **question**(s) that the author is addressing in the textbook is(are)_

_____.

3) The most important kinds of **information** in this textbook are _____

_____.

4) The main **inferences** (and conclusions) in this textbook are _____

_____.

5) The key **concept**(s) we need to understand in this textbook is(are) _____

_____.

By these concepts the author means_____

_____.

6) The main **assumption**(s) underlying the author's thinking is(are) _____

_____.

7a) If people take the textbook seriously, the **implications** are _____

_____.

7b) If people fail to take the textbook seriously, the **implications** are _____

_____.

8) The main **point(s) of view** presented in this textbook is(are)_____

_____.

Experimental Thinking
Requires Experimental Controls

To maintain control over all likely casual factors being examined, experimenters isolate each variable and observe its effects on the phenomena being studied to determine which factors are essential to the causal effects.

Experiments Can Go Awry When Scientists Fail to Control for Confounded Variables. Often, a range of variables are 'associated' with a given effect, while only one of the variables is truly responsible for the effect. For example, it has been found that in France, where people drink a lot of red wine, the incidence of heart attacks is lower than in countries of northern Europe where red wine is less popular. Can we conclude from this statistical study that the regular drinking of moderate amounts of red wine can prevent the occurrence of heart attacks? No, because there are many other differences between the life styles of people in France and those in northern Europe, for example diet, work habits, climate, smoking, commuting, air pollution, inherited pre-dispositions, etc. These other variables are 'associated' or 'confounded' with the red wine variable. One or more of these confounded variables might be the actual cause of the low incidence of heart attacks in France. These variables would have to be controlled in some way before one could conclude that drinking red wine lowers the incidence of heart attacks.

A possible experimental design would be to compare Frenchmen who drink red wine with those who drink no alcohol at all or drink beer — making sure that these groups do not differ on any other measurable variables. Or we might study northern Europeans who drink red wine and see if the incidence of heart attack is lower among them than among northern Europeans who do not drink red wine. We could also take a group of patients who have had a heart attack, and instruct one half to drink a little red wine every day, and tell the other group to drink apple juice. After a number of years we could compare the rate of incidence of heart attacks in the two groups.

The Logic of an Experiment

(Attach a detailed description of the experiment or laboratory procedure.)

The main goal of the experiment is _____

_____ .

The hypothesis(es) we seek to test in this experiment is(are) _____

_____

_____ .

The key question the experiment seeks to answer is _____

_____ .

The controls involved in this experiment are _____

_____ .

The key concept(s) or theory(ies) behind the experiment is(are) _____

_____

_____ .

The experiment is based on the following assumptions_____

_____

_____ .

The data that will be collected in the experiment are_____

_____

_____ .

The potential implications of the experiment are _____

_____ .

The point of view behind the experiment is _____

_____ .

Post Experiment Analysis

The data collected during the experiment was _____

_____

_____

_____

_____.

The inferences (conclusions) that most logically follow from the data are _____

_____

_____

_____.

These inferences are/are not debatable, given the data gathered in this study and the evidence to this point.

The hypothesis (or hypotheses) for this experiment was/was not (were/were not) support by the experiment results.

The assumptions made prior to this experiment should/should not be modified given the data gathered in this experiment. Modifications to assumptions (if any) should be as follows_____

_____

_____

_____.

The most significant implications of this experiment are_____

_____

_____

_____.

Recommendations for future research in this area are_____

_____

_____

_____.

How to Evaluate an Author's or Experimenter's Scientific Reasoning

1. Focusing on the stated scientific **Purpose:** Is the purpose of the author well-stated or clearly implied? Is it justifiable?

2. Focusing on the key scientific **Question:** Is the question at issue well-stated (or clearly implied)? Is it clear and unbiased? Does the expression of the question do justice to the complexity of the matter at issue? Are the question and purpose directly relevant to each other?

3. Focusing on the most important scientific **Information** or data: Does the writer cite relevant evidence, experiences, and/or information essential to the issue? Is the information accurate and directly relevant to the question at issue? Does the writer address the complexities of the issue? Does the experimenter clearly delineate the scientific data to be collected?

4. Focusing on the most fundamental **Concepts** at the heart of the scientific reasoning: Are the key ideas clarified? Are the ideas used justifiably? Does the experimenter clarify the theories behind the experiment?

5. Focusing on **Assumptions:** Does the scientific reasoner clearly delineate the scientific assumptions? Does s/he show a sensitivity to what s/he is taking for granted or assuming (insofar as those assumptions might reasonably be questioned)? Or does the reasoner use questionable assumptions without addressing problems inherent in those assumptions?

6. Focusing on the most important scientific **Inferences** or conclusions: Do the inferences and conclusions made by the scientific reasoner clearly follow from the information relevant to the issue, or does the reasoner jump to unjustifiable conclusions? Does the reasoner consider alternative conclusions where the scientific issue is complex? In other words, does the reasoner use a sound line of reasoning to come to logical scientific conclusions, or can you identify flaws in the reasoning somewhere? Does the experimenter clearly separate data from conclusions?

7. Focusing on the scientific **Point of View:** Does the reasoner show a sensitivity to alternative relevant scientific points of view or lines of reasoning? Does s/he consider and respond to objections framed from other relevant scientific points of view?

8. Focusing on **Implications:** Does the reasoner display a sensitivity to the implications and consequences of the position s/he is taking?

Two Kinds of Scientific Questions

In approaching a question, it is useful to figure out what type it is. Is it a question with one definitive answer? Or does the question require us to consider competing points of view?

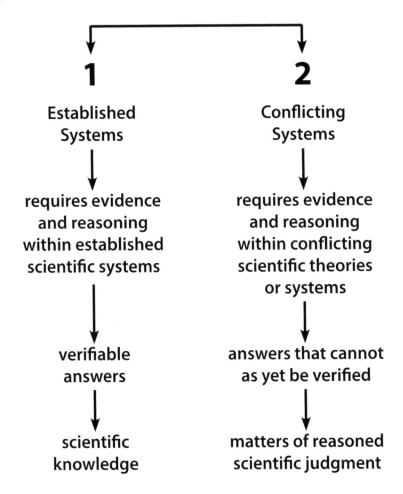

1

Established Systems

requires evidence and reasoning within established scientific systems

verifiable answers

scientific knowledge

2

Conflicting Systems

requires evidence and reasoning within conflicting scientific theories or systems

answers that cannot as yet be verified

matters of reasoned scientific judgment

See explications and examples of both types of questions on the following two pages.

Asking One System and Conflicting System Questions

There are a number of essential ways to categorize questions for the purpose of analysis. One such way is to focus on the type of reasoning required by the question. With **one system** questions, there is an established procedure or method for finding the answer. With **conflicting system** questions, there are multiple competing viewpoints from which, and within which, one might reasonably pursue an answer to the question. There are better and worse answers, but no verifiable "correct" ones, since these are matters about which even experts disagree (hence the "conflict" from system to system).

Questions of Procedure (established- or one-system) – These include questions with an established procedure or method for finding the answer. These questions are settled by facts, by definition, or both. They are prominent in science and mathematics.

Examples of one-system (monological) scientific questions:

- What is science?
- What is biology?
- What are some methods scientists use in making discoveries and developing theories?
- How do these methods differ from study in "non-scientific" fields?
- What kind of systematic study is characteristic of science?
- What significant positive implications have resulted from scientific research?
- What roles do math and logic play in scientific thinking?
- What is the boiling point of lead?
- What are the main branches of science and how do they interrelate?
- What is botany?
- What is plant classification and why is it important?
- How do plants function, both as a group and individually?
- What are some important uses of plant life, in medicine, in lumber production, in food production?

Questions of Judgment (conflicting systems) – Questions requiring reasoning, but with more than one arguable answer. These are questions that make sense to debate, questions with better-or-worse answers (well-supported and reasoned or poorly-supported and/or poorly-reasoned). Here we are seeking the best answer within a range of possibilities. We evaluate answers to these questions using universal intellectual standards such as clarity, accuracy, relevance, etc. These questions are predominant in the human disciplines (history, philosophy, economics, sociology, art…). But there are many such questions in the sciences as well.

Examples of conflicting system (multilogical) scientific questions:

- What are the most significant ways that scientific research can be misused?
- How can we balance business interests and ecological preservation?
- How is scientific thinking making a contribution to our personal lives? Are there any ways in which it is a threat? If so, what?
- What are the most significant limitations of science?
- How can we balance exploitation of plant life for the needs and desires of people with maintaining essential plant life on earth?
- What is the best system we can construct for analyzing, classifying, and understanding all forms of animal life?
- What are the most significant barriers to the application of science in human life?
- Do the risks outweigh the benefits in taking bio-identical hormones?
- To what extent is psychology scientific? To what extent is it not?

Scientific Reasoning Abilities

Scientific Affective Dimensions

- exercising independent thought and judgment
- developing insight into egocentrism and sociocentrism
- exercising reciprocity
- suspending judgment

Cognitive Dimensions: Scientific Macro-Abilities

- avoiding oversimplification of scientific issues
- developing scientific perspective
- clarifying scientific issues and claims
- clarifying scientific ideas
- developing criteria for scientific evaluation
- evaluating scientific authorities
- raising and pursuing root scientific questions
- evaluating scientific arguments
- generating and assessing solutions to scientific problems
- identifying and clarifying scientific points of view
- engaging in Socratic discussion on scientific issues
- practicing dialectical thinking on scientific issues

Cognitive Dimensions: Scientific Micro-Skills

- distinguishing scientific facts from scientific principles, values, and ideas
- evaluating assumptions
- distinguishing scientifically relevant from scientifically irrelevant facts
- making plausible scientific inferences
- supplying evidence for a scientific conclusion
- recognizing contradictions
- recognizing scientific implications and consequences
- refining scientific generalizations

Analyzing & Assessing Scientific Research

Use this template to assess the quality of any scientific research project or paper.

1) All scientific research has a fundamental PURPOSE and goal.
   - Research purposes and goals should be clearly stated.
   - Related purposes should be explicitly distinguished.
   - All segments of the research should be relevant to the purpose.
   - All research purposes should be realistic and significant.
2) All scientific research addresses a fundamental QUESTION, problem or issue.
   - The fundamental question at issue should be clearly and precisely stated.
   - Related questions should be articulated and distinguished.
   - All segments of the research should be relevant to the central question.
   - All research questions should be realistic and significant.
   - All research questions should define clearly stated intellectual tasks that, being fulfilled, settle the questions.
3) All scientific research identifies data, INFORMATION, and evidence relevant to its fundamental question and purpose.
   - All information used should be clear, accurate, and relevant to the fundamental question at issue.
   - Information gathered must be sufficient to settle the question at issue.
   - Information contrary to the main conclusions of the research should be explained.
4) All scientific research contains INFERENCES or interpretations by which conclusions are drawn.
   - All conclusions should be clear, accurate, and relevant to the key question at issue.
   - Conclusions drawn should not go beyond what the data imply.
   - Conclusions should be consistent and reconcile discrepancies in the data.
   - Conclusions should explain how the key questions at issue have been settled.
5) All scientific research is conducted from some POINT OF VIEW or frame of reference.
   - All points of view in the research should be identified.
   - Objections from competing points of view should be identified and fairly addressed.
6) All scientific research is based on ASSUMPTIONS.
   - Clearly identify and assess major assumptions in the research.
   - Explain how the assumptions shape the research point of view.
7) All scientific research is expressed through, and shaped by, CONCEPTS and ideas.
   - Assess for clarity the key concepts in the research.
   - Assess the significance of the key concepts in the research.
8) All scientific research leads somewhere (i.e., have IMPLICATIONS and consequences).
   - Trace the implications and consequences that follow from the research.
   - Search for negative as well as positive implications.
   - Consider all significant implications and consequences.

Purpose

(All scientific reasoning has a purpose.)

Primary Standards: (1) Clarity, (2) Significance, (3) Achievability
(4) Consistency, (5) Justifiability

Common Problems: (1) Unclear, (2) Trivial, (3) Unrealistic, (4) Contradictory,
(5) Unfair

Principle: To reason well, you must clearly understand your purpose, and
your purpose must be reasonable and fair.

Skilled Thinkers...	Unskilled Thinkers...	Critical Reflections
Take the time to state their purpose clearly.	Are often unclear about their central purpose.	Have I made the purpose of my reasoning clear? What exactly am I trying to achieve? Have I stated the purpose in several ways to clarify it?
Distinguish it from related purposes.	Oscillate between different, sometimes contradictory purposes.	What different purposes do I have in mind? How do I see them as related? Am I going off in somewhat different directions? How can I reconcile these contradictory purposes?
Periodically remind themselves of their purpose to determine whether they are straying from it.	Lose track of their fundamental object or goal	In writing this proposal, do I seem to be wandering from my purpose? How do my third and fourth paragraph relate to my central goal?
Adopt realistic purposes and goals.	Adopt unrealistic purposes and set unrealistic goals.	Am I trying to accomplish too much in this project?
Choose significant purposes and goals.	Adopt trivial purposes and goals as if they were significant.	What is the significance of pursuing this particular purpose? Is there a more significant purpose I should be focused on?
Choose goals and purposes that are consistent with other goals and purposes they have chosen.	Inadvertently negate their own purposes. Do not monitor their thinking for inconsistent goals.	Does one part of my proposal seem to undermine what I am trying to accomplish in another part?
Adjust their thinking regularly to their purpose.	Do not adjust their thinking regularly to their purpose.	Does my argument stick to the issue? Am I acting consistently within my purpose?
Choose purposes that are fair-minded, considering the desires and rights of others equally with their own desires and rights.	Choose purposes that are self-serving at the expense of others' needs and desires.	Is my purpose self-serving or concerned only with my own desires? Does it take into account the rights and needs of other people?

Questions at Issue or Central Problem

(All scientific reasoning is an attempt to figure something out,
to settle some question, solve some problem.)

Primary Standards: (1) Clarity and precision, (2) Significance, (3) Answerability (4) Relevance

Common Problems: (1) Unclear and imprecise, (2) Insignificant, (3) Not answerable, (4) Irrelevant

Principle: To settle a question, it must be answerable, and you must be clear about it and understand what is needed to adequately answer it..

Skilled Thinkers...	Unskilled Thinkers...	Critical Reflections
Are clear about the question they are trying to settle.	Are often unclear about the question they are asking.	Am I clear about the main question at issue? Am I able to state it precisely?
Can re-express a question in a variety of ways.	Express questions vaguely and find questions difficult to reformulate for clarity.	Am I able to reformulate my question in several ways to recognize the complexity of it?
Can break a question into sub-questions.	Are unable to break down the questions they are asking	Have I broken down the main question into sub-questions? What are the sub-questions embedded in the main question?
Routinely distinguish questions of different types.	Confuse questions of different types and thus often respond inappropriately to the questions they ask.	Am I confused about the type of question I am asking? For example: Am I confusing a legal question with an ethical one? Am I confusing a question of preference with a question requiring judgment?
Distinguish significant from trivial questions.	Confuse trivial questions with significant ones.	Am I focusing on trivial questions while other significant questions need to be addressed?
Distinguish relevant questions from irrelevant ones.	Confuse irrelevant questions with relevant ones.	Are the questions I am raising in this discussion relevant to the main question at issue?
Are sensitive to the assumptions built into the questions they ask.	Often ask loaded questions.	I the way I am putting the questions loaded? Am I taking for granted from the onset the correctness of my own position?
Distinguish questions they can answer from questions they can't.	Try to answer questions they are not in a position to answer.	Am I in a position to answer this question? What information would I need to have before I could answer the question?

Information

(All scientific reasoning is based on data, information,
evidence, experience, and research.)

Primary Standards: (1) Clear, (2) Relevant, (3) Fairly gathered and
reported, (4) Accurate, (5) Adequate, (6) Consistently
applied

Common Problems: (1) Unclear, (2) Irrelevant, (3) Biased, (4) Inaccurate, (5)
Insufficient, (6) Inconsistently applied

Principle: Reasoning can be only as sound as the information upon which it
is based..

Skilled Thinkers...	Unskilled Thinkers...	Critical Reflections
Assert a claim only when they have sufficient evidence to back it up.	Assert claims without considering all relevant information.	Is my assertion supported by evidence?
Can articulate and evaluate the information behind their claims.	Do not articulate the information they are using in their reasoning and so do not subject it to rational scrutiny.	Do I have evidence to support my claim that I have not clearly articulated? Have I evaluated for accuracy and relevance the information I am using?
Actively search for information against (not just for) their position.	Gather information only when it supports their point of view.	Where is a good place to look for evidence on the opposite side? Have I looked there? Have I honestly considered information that does not support my position?
Focus on relevant information and disregard what is irrelevant to the question at issue.	Do not carefully distinguish between relevant information and irrelevant information.	Are my data relevant to the claim I am making? Have I failed to consider relevant information?
Draw conclusions only to the extent that they are supported by the data and sound reasoning.	Make inferences that go beyond what the data supports.	Does my claim go beyond the evidence I have cited?
State their evidence clearly and fairly.	Distort the data or state it inaccurately.	Is my presentation of the pertinent information clear and coherent? Have I distorted information to support my position?

Inference and Interpretation

(All scientific reasoning contains inferences from which
we draw conclusions and give meaning to data and situations.)

Primary Standards: (1) Clarity, (2) Logicality, (3) Justifiability, (4) Profundity,
(5) Reasonability, (6) Consistency

Common Problems: (1) Unclear, (2) Illogical, (3) Unjustified, (4) Superficial,
(5) Unreasonable, (6) Contradictory

Principle: Reasoning can be only as sound as the inferences it makes (or the
conclusions to which it comes)

Skilled Thinkers...	Unskilled Thinkers...	Critical Reflections
Are clear about the inferences they are making. Clearly articulate their inferences.	Are often unclear about the inferences they are making. Do not clearly articulate their inferences.	Am I clear about the inferences I am making? Have I clearly articulated my conclusions?
Usually make inferences that follow from the evidence or reasons presented.	Often make inferences that do not follow from the evidence or reasons presented.	Do my conclusions logically follow from the evidence and reasons presented?
Often make inferences that are deep rather than superficial.	Often make inferences that are superficial.	Are my conclusions superficial, given the problem?
Often make inferences or come to conclusions that are reasonable.	Often make inferences or come to conclusions that are unreasonable.	Are my conclusions unreasonable?
Make inferences or come to conclusions that are consistent with each other.	Often make inferences or come to conclusions that are contradictory.	Do the conclusions I reach in the first part of my analysis seem to contradict the conclusions that I come to at the end?
Understand the assumptions that lead to inferences.	Do not seek to figure out the assumptions that lead to inferences.	Is my inference based on a faulty assumption? How would my inference be changed if I were to base it on a different, more justifiable assumption?

Assumptions

(All scientific reasoning is based on assumptions—beliefs we take for granted.)

Primary Standards: (1) Clarity, (2) Justifiability, (3) Consistency
Common Problems: (1) Unclear, (2) Unjustified, (3) Contradictory
Principle: Reasoning can be only as sound as the assumptions on which it is based.

Skilled Thinkers...	Unskilled Thinkers...	Critical Reflections
Are clear about the assumptions they are making.	Are often unclear about the assumptions they make.	Are my assumptions clear to me? Do I clearly understand what my assumptions are based on?
Make assumptions that are reasonable and justifiable given the situation and evidence.	Often make unjustified or unreasonable assumptions.	Do I make assumptions about the future based on just one experience from the past? Can I fully justify what I am taking for granted? Are my assumptions justifiable given the evidence I am using to support them?
Make assumptions that are consistent with each other.	Make assumptions that are contradictory.	Do the assumptions I made in the first part of my argument contradict the assumptions I am making now?
Constantly seek to discern and understand their assumptions.	Ignore their assumptions.	What assumptions am I making in this situation? Are they justifiable? Where did I get these assumptions?

Concepts and Ideas

(All scientific reasoning is expressed through,
and shaped by, concepts and ideas.)

Primary Standards: (1) Clarity, (2) Relevancy, (3) Depth, (4) Accuracy
Common Problems: (1) Unclear, (2) Irrelevant, (3) Superficial, (4) Inaccurate
Principle: Reasoning can be only as sound as the assumptions on which it is based.

Skilled Thinkers...	Unskilled Thinkers...	Critical Reflections
Recognize the key concepts and ideas they and others use.	Are unaware of the key concepts and ideas they and others use.	What is the main concept I am using in my thinking? What are the main concepts others are using?
Are able to explain the basic implications of the key words and phrases they use.	Cannot accurately explain basic implications of their key words and phrases.	Am I clear about the implications of key concepts? For example: Does the word "argument" have negative implications that the word "rationale" does not?
Distinguish special, nonstandard uses of words from standard uses, and avoid jargon in inappropriate settings.	Do not recognize when their use of a word or phrase or symbol departs from conventional or disciplinary usage.	Where did I get my definitions of this central concept? Is it consistent with convention? Have I put unwarranted conclusions into the definition? Does any of my vocabulary have special connotations that others may not recognize? Have I been careful to define any specialized terms, abbreviations, or mathematical symbols? Have I avoided jargon where possible?
Recognize irrelevant concepts and ideas and use concepts and ideas in ways relevant to their functions.	Use concepts or theories in ways inappropriate to the subject or issue.	Am I using the concept of "efficiency" appropriately? For example: Have I confused "efficiency" and "effectiveness"? Am I applying theories which do not apply to this application?
Think deeply about the concepts they use.	Fail to think deeply about the concepts they use.	Am I thinking deeply enough about this concept? For example: The concept of product safety or durability, as I describe it, does not take into account inexpert customers. Do I need to consider the idea of product safety more deeply?

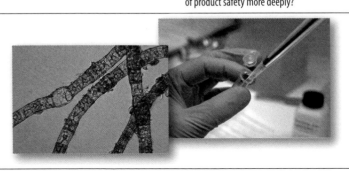

Point of View

(All scientific reasoning is done from some point of view.)

Primary Standards: (1) Flexibility, (2) Fairness, (3) Clarity, (4) Breadth, (5) Relevance

Common Problems: (1) Restricted, (2) Biased, (3) Unclear, (4) Narrow, (5) Irrelevant

Principle: To reason well, you must identify those points of view relevant to the issue and enter these viewpoints empathetically.

Skilled Thinkers...	Unskilled Thinkers...	Critical Reflections
Keep in mind that people have different points of view, especially on controversial issues.	Dismiss or disregard alternative reasonable viewpoints.	Have I articulated the point of view from which I am approaching this issue? Have I considered opposing points of view regarding this issue?
Consistently articulate other points of view and reason from within those points of view to adequately understand other points of view.	Cannot see issues from points of view that are significantly different from their own. Cannot reason with empathy from alien points of view.	I may have characterized my own point of view, but have I considered the most significant aspects of the problem from the point of view of others?
Seek other viewpoints, especially when the issue is one they believe in passionately.	Recognize other points of view when the issue is not emotionally charged, but cannot do so for issues about which they feel strongly.	Am I expressing X's point of view in an unfair manner? Am I having difficulty appreciating X's viewpoint because I am emotional about this issue?
Confine their monological reasoning to problems that are clearly monological.*	Confuse multilogical with monological issues; insists that there is only one frame of reference within which a given multilogical question must be decided.	Is the question here monological or multilogical? How can I tell? Am I reasoning as if only one point of view is relevant to this issue when in reality other viewpoints are relevant?
Recognize when they are most likely to be prejudiced.	Are unaware of their own prejudices.	Is this prejudiced or reasoned judgment? If prejudiced, where does it originate?
Approach problems and issues with a richness of vision and an appropriately broad point of view.	Reason from within inappropriately narrow or superficial points of view.	Is my approach to this question too narrow? Am I considering other viewpoints so I can adequately address the problem?

*Monological problems are ones for which there are definite correct and incorrect answers and definite procedures for getting those answers. In multilogical problems, there are competing schools of thought to be considered.

Implications and Consequences

(All scientific reasoning leads somewhere. It has implications and,
when acted upon, has consequences.)

Primary Standards: (1) Significance, (2) Logicality, (3) Clarity, (4) Precision, (5) Completeness

Common Problems: (1) Unimportant, (2) Unrealistic, (3) Unclear, (4) Imprecise, (5) Incomplete

Principle: To reason well through an issue, you might think through the implications that follow from your reasoning. You must think through the consequences likely to flow from the decisions you make.

Skilled Thinkers...	Unskilled Thinkers...	Critical Reflections
Trace out a number of significant potential implications and consequences of their reasoning.	Trace out few or none of the implications and consequences of holding a position or making a decision.	Did I spell out all the significant consequences of the action I am advocating? If I were to take this course of action, what other consequences might follow that I have not considered? Have I considered all plausible failures?
Clearly and precisely articulate the possible implications and consequences.	Are unclear and imprecise in the possible consequences they articulate.	Have I delineated clearly and precisely the consequences likely to follow from my chosen actions?
Search for potentially negative as well as potentially positive consequences.	Trace out only the consequence they had in mind at the beginning, either positive or negative, but usually not both.	I may have done a good job of spelling out some positive implications of the decision I am about to make, but what are some of the possible negative implications or consequences.
Anticipate the likelihood of unexpected negative and positive implications.	Are surprised when their decisions have unexpected consequences.	If I make this decision, what are some possible unexpected implications? What are some of the variables out of my control that might lead to negative consequences?
Considers the reactions of all parties.	Assumes the outcomes and products will be welcomed by other parties.	What measures are appropriate to inform the community or marketplace? What opinion leaders should be involved?

Intellectual Dispositions
Essential to Scientific Thinking

To become fair-minded, intellectually responsible scientific thinkers, we must develop intellectual virtues or dispositions. These attributes are essential to excellence of scientific thought.

They determine with what insight and integrity we think. This section contains brief descriptions of the intellectual virtues, along with related questions that foster their development.

Intellectual humility is knowledge of ignorance, sensitivity to what you know and what you do not know. It implies being aware of your biases, prejudices, self-deceptive tendencies and the limitations of your viewpoint. Questions that foster intellectual humility in scientific thinking include:

- What do I really know about the scientific issue I am raising?
- To what extent do my prejudices or biases influence my ability to think scientifically?
- How do the beliefs I have uncritically accepted keep me from thinking scientifically?

Intellectual courage is the disposition to question beliefs you feel strongly about. It includes questioning the beliefs of your culture and the groups to which you belong, and a willingness to express your views even when they are unpopular. Questions that foster intellectual courage include:

- To what extent have I analyzed the beliefs I hold which may impede my ability to think scientifically?
- To what extent have I demonstrated a willingness to give up my beliefs when sufficient scientific evidence is presented against them?
- To what extent am I willing to stand up against the majority (even though people ridicule me)?

Intellectual empathy is awareness of the need to actively entertain views that differ from our own, especially those we strongly disagree with. It is to accurately reconstruct the viewpoints and reasoning of our opponents and to reason from premises, assumptions, and ideas other than our own. Questions that foster intellectual empathy include:

- To what extent do I accurately represent scientific viewpoints I disagree with?
- Can I summarize the scientific views of my opponents to their satisfaction?
- Can I see insights in the scientific views of others and prejudices in my own?

Intellectual integrity consists in holding yourself to the same intellectual standards you expect others to honor (no double standards). Questions that foster intellectual integrity in scientific thinking include:

- To what extent do I expect of myself what I expect of others?
- To what extent are there contradictions or inconsistencies in the way I deal with scientific issues?
- To what extent do I strive to recognize and eliminate self-deception when reasoning through scientific issues?

Intellectual perseverance is the disposition to work your way through intellectual complexities despite the frustration inherent in the task. Questions that foster intellectual perseverance in scientific thinking include:

- Am I willing to work my way through complexities in a scientific issue or do I tend to give up when I experience difficulty?
- Can I think of a difficult scientific problem concerning which I have demonstrated patience and determination in working through its difficulties?

Confidence in reason is based on the belief that one's own higher interests and those of humankind at large are best served by giving the freest play to reason. It means using standards of reasonability as the fundamental criteria by which to judge whether to accept or reject any belief or position. Questions that foster confidence in reason when thinking scientifically include:

- Am I willing to change my position when the scientific evidence leads to a more reasonable position?
- Do I adhere to scientific principles and evidence when persuading others of my position or do I distort matters to support my position?
- Do I encourage others to come to their own scientific conclusions or do I try to force my views on them?

Intellectual autonomy is thinking for oneself while adhering to standards of rationality. It means thinking through issues using one's own thinking rather than uncritically accepting the viewpoints of others. Questions that foster intellectual autonomy in scientific thinking include:

- Do I think through scientific issues on my own or do I merely accept the scientific views of others?
- Having thought through a scientific issue from a rational perspective, am I willing to stand alone despite the irrational criticisms of others?

Scientific Thinkers Routinely
Apply the Intellectual Standards
to the Elements of Scientific Reasoning
as they develop the traits of a scientific mind

ESSENTIAL INTELLECTUAL STANDARDS

Clarity	Precision
Accuracy	Significance
Relevance	Completeness
Logicalness	Fairness
Breadth	Depth

Must be applied to

THE ELEMENTS OF SCIENTIFIC THOUGHT

Purposes	Inferences
Questions	Concepts
Points of view	Implications
Information	Assumptions

As we learn to develop

THE TRAITS OF A SCIENTIFIC MIND

Intellectual Humility	Intellectual Perseverance
Intellectual Autonomy	Confidence in Reason
Intellectual Integrity	Intellectual Empathy
Intellectual Courage	Fairmindedness

Development of the Scientific Mind

**Accomplished
Scientific Thinker**
(Good habits of scientific
thought are becoming
second nature)

**Advanced
Scientific Thinker**
(We advance in keeping with
our practice)

**Practicing
Scientific Thinker**
(We recognize the need for
regular practice)

**Beginning
Scientific Thinker**
(We try to improve our
scientific thinking, but without
regular practice)

Challenged Thinker
(We begin to recognize the fact
that we often fail to think
scientifically when dealing with
scientific questions)

Unreflective Thinker
(We are unaware of significant problems
in our thinking about scientific issues,
hence we are unable to distinguish science
from pseudo-science)

Analyzing the Logic of a Subject

When we understand the elements of reasoning, we realize that all subjects, all disciplines, have a fundamental logic defined by the structures of thought embedded in them.

Therefore, to lay bare a subject's most fundamental logic, we should begin with these questions:

- What is the main <u>purpose</u> or <u>goal</u> of studying this subject? What are people in this field trying to accomplish?
- What kinds of <u>questions</u> do they ask? What kinds of problems do they try to solve?
- What sorts of <u>information</u> or data do they gather?
- What types of <u>inferences</u> or judgments do they typically make? (Judgments about…)
- How do they go about gathering information in ways that are distinctive to this field?
- What are the most basic ideas, <u>concepts</u> or theories in this field?
- What do professionals in this field take for granted or <u>assume</u>?
- How should studying this field affect my view of the world?
- What <u>viewpoint</u> is fostered in this field?
- What <u>implications</u> follow from studying this discipline? How are the products of this field used in everyday life?

These questions can be contextualized for any given class day, chapter in the textbook and dimension of study. For example, on any given day you might ask one or more of the following questions:

- What is our main <u>purpose</u> or <u>goal</u> today? What are we trying to accomplish?
- What kinds of <u>questions</u> are we asking? What kinds of problems are we trying to solve? How does this problem relate to everyday life?
- What sort of <u>information</u> or data do we need? How can we get that information?
- What is the most basic idea, <u>concept</u> or theory we need to understand to solve the problem we are most immediately posing?
- From what <u>point of view</u> should we look at this problem?
- What can we safely <u>assume</u> as we reason through this problem?
- Should we call into question any of the <u>inferences</u> that have been made?
- What are the <u>implications</u> of what we are studying?

The Logic of Scientific Reasoning

Point of View
Looking at the physical world as something to be understood through careful observation and systematic study

Purpose
To figure out how the physical world operates through systematic observation and experimentation

Question
What can be figured out about how the physical world operates by observation and experimentation

Implications and Consequences
If we systematically study the physical world, we can gain important knowledge about that world.

Elements of Reasoning

Information
Facts that can be systematically gathered about the physical world

Assumptions
That there are laws at work in the physical world that can be figured out through systematic observation and experimentation

Essential Concepts
The workings of the physical world as predictable and understandable through carefully designed hypotheses, predictions and experimentation

Interpretation and Inference
Judgements based on observations and experimentation that lead to systematized knowledge of nature and the physical world

The Questioning Mind in Science
Newton, Darwin, and Einstein[1]

Most people think that genius is the primary determinant of intellectual achievement. Yet three of the most distinguished thinkers had in common, not inexplicable genius, but a questioning mind. Their intellectual skills and inquisitive drive embodied the essence of critical thinking. Through skilled deep and persistent questioning they redesigned our view of the physical world and the universe.

Consider Newton. Uninterested in the set curriculum at Cambridge, Newton at 19 drew up a list of questions under 45 heads. His title: "Quaestiones," signaled his goal: constantly to question the nature of matter, place, time, and motion. His style was to slog his way to knowledge. For example, he "bought Descartes's Geometry and read it by himself. When he got over two or three pages he could understand no farther, then he began again and advanced farther and continued so doing till he made himself master of the whole…"

When asked how he had discovered the law of universal gravitation, he said: "By thinking on it continually." I keep the subject constantly before me and wait till the first dawnings open slowly, by little and little, into a full and clear light." This pattern of consistent, almost relentless questioning, led to depth of understanding and reconstruction of previous theories about the universe.

Newton acutely recognized knowledge as a vast field to be discovered: "I don't know what I may seem to the world, but, as to myself, I seem to have been only like a boy playing on the sea shore, and diverting myself in now and then finding a smoother pebble or prettier shell than ordinary, whilst the great ocean of truth lay all undiscovered before me."

---

1   (Newton: The Life of Isaac Newton, by Richard Westfall, NY: Cambridge University Press,1993; The Autobiography of Charles Darwin, by Francis Darwin, NY: Dover Publications, 1958; A. Einstein: The Life and Times, by Ronald Clark, NY: Avon Books, 1984; A Variety of Men, by C.P. Snow, NY: Charles Scribners and Sons, 1967).)

Darwin's experience and approach to learning were similar to Newton's. First, he found traditional instruction discouraging. "During my second year at Edinburgh I attended lectures on Geology and Zoology, but they were incredibly dull. The sole effect they produced in me was the determination never as long as I lived to read a book on Geology, or in any way to study the science."

His experience at Cambridge was similar: "During the three years which I spent at Cambridge my time was wasted… The work was repugnant to me, chiefly from my not being able to see any meaning in [it]…"

Like Newton and Einstein, Darwin had a careful mind rather than a quick one:

"I have as much difficulty as ever in expressing myself clearly and concisely; and this difficulty has caused me a very great loss of time, but it has had the compensating advantage of forcing me to think long and intently about every sentence, and thus I have been led to see errors in reasoning and in my own observations or those of others."

In pursuing intellectual questions, Darwin relied upon perseverance and continual reflection, rather than memory and quick reflexes. "I have never been able to remember for more than a few days a single date or line of poetry." Instead, he had, "the patience to reflect or ponder for any number of years over any unexplained problem…At no time am I a quick thinker or writer: whatever I have done in science has solely been by long pondering, patience, and industry."

Einstein, for his part, did so poorly in school that when his father asked his son's headmaster what profession his son should adopt, the answer was simply, "It doesn't matter; he'll never make a success of anything." In high school, the regimentation "created in him a deep suspicion of authority. This feeling lasted all his life, without qualification."

Einstein commented that his schooling required "the obedience of a corpse." The effect of the regimented school was a clear-cut reaction by Einstein; he learned "to question and doubt." He concluded: "…youth is intentionally being deceived by the state through lies."

He showed no signs of being a genius, and as an adult denied that his mind was extraordinary: "I have no particular talent. I am merely extremely inquisitive." He failed his entrance examination to the Zurich Polytechnic. When he finally passed, "the examinations so constrained his mind that, when he had graduated, he did not want to think about scientific problems for a year." His final exam was so non-distinguished that afterward he was refused a post as an assistant (the lowest grade of postgraduate job).

Exam-taking, then, was not his forte. Questioning deeply and thinking critically was.

Einstein had the basic critical thinking ability to cut problems down to size: "one of his greatest intellectual gifts, in small matters as well as great, was to strip off the irrelevant frills from a problem."

When we consider the work of these three thinkers, Einstein, Darwin, and Newton, we find, not the unfathomable, genius mind. Rather we find thinkers who placed deep and fundamental questions at the heart of their work and pursued them passionately.

The Logic of Science

Goals Scientists Pursue: Scientists seek to figure out how the physical world operates through systematic observation, experimentation, and analysis. By analyzing the physical world, they seek to formulate principles, laws, and theories useful in explaining natural phenomena, and in guiding further scientific study.

Questions Scientists Ask: How does the physical world operate? What are the best methods for figuring things out about the physical world? What are the barriers to figuring things out about the physical world? How can we overcome those barriers?

Information Scientists Use: Scientists as a whole use virtually any type of information that can be gathered systematically through observation and measurement, though most specialize in analyzing specific kinds of information. To name just some of the information scientists use, they observe and examine plants, animals, planets, stars, rocks, rock formations, minerals, bodies of water, fossils, chemicals, phenomena in the earth's atmosphere and cells. They also observe interactions between phenomena.

Judgments Scientists Make: Scientists make judgments about the physical world based on observations and experimentation. These judgments lead to systematized knowledge, theories, and principles helpful in explaining and understanding the world.

Concepts that Guide Scientists' Thinking: The most fundamental concepts that guide the thinking of scientists are 1) physical world (of nature and all matter); 2) hypothesis (an unproved theory, proposition, or supposition tentatively accepted to explain certain facts or to provide a basis for further investigation); 3) experimentation (a systematic and operationalized process designed to figure out something about the physical world); and 4) systematic observation (the act or practice of noting or recording facts or events in the physical world). Other fundamental concepts in science include: theory, law, scientific method, pure sciences, and applied sciences.

Key Assumptions Scientists Make: 1) There are laws at work in the physical world that can be figured out through systematic observation and experimentation; 2) Much about the physical world is still unknown; 3) Through science, the quality of life on earth can be enhanced.

Implications of Science: Many important implications and consequences have resulted from scientific thinking, some of which have vastly improved the quality of life on earth, others of which have resulted in decreased quality of life (e.g., the destruction of the earth's forests, oceans, natural habitats, etc.). One important positive implication of scientific thinking is that it enables us to replace mythological thinking with theories and principles based in scientific fact.

The Scientific Point of View: Scientists look at the physical world and see phenomena best understood through careful observation and systematic study. They see scientific study as vital to understanding the physical world and replacing myth with scientific knowledge.

The Logic of Physics

The Goal of Physics is to discover the physical forces, interactions, and properties of matter, including the physical properties of the atom and sub-atomic particles. In pursuing this end, physicists study gravitation, motion, space, time, force, and energy. This entails the study of mechanics, heat, light, sound, electricity, magnetism, and the constitution of matter. Physics conducts its study of the physical properties of matter and energy insofar as these properties can be measured, expressed in mathematical formulas, and explained by physical theories. Its goals may be contrasted with those of chemistry (which focuses on chemical properties, on the composition and transformations of matter) and those of biology (which focuses on living matter).

Its Key Question is: What are the physical properties of matter and energy insofar as both can be measured, expressed in mathematical formulas, and explained by physical theories? (Physical properties can change without changing the identity of the matter; chemical properties cannot change without changing the identity of the matter.)

Its Key Concepts include: matter, energy, mass, space, time, light, work, entropy, motion, volume, density, weight, magnitude, direction, displacement, velocity, acceleration, momentum, inertia, equilibrium, friction, gravitation, mechanics, heat, sound, electricity, magnetism, chaos theory, quantum, and relativity.

Its Key Assumptions are: that the universe is controlled by laws, that the same laws apply throughout the universe, that the laws guiding the universe can be expressed in mathematical terms and formulas, that physical properties can be distinguished from chemical ones, that the velocity of light is constant throughout, that space and time are interrelated, that all motion is relative, and that the forces of inertia, gravitation, and electromagnetism are different manifestations of a single force.

The Data or Information Physicists Gather are all focused on the causal relations or statistical correlations of physical occurrences or phenomena. Physicists use information from many physical sources such as heat, light, sound, mechanics, electricity, and magnetism to come to conclusions about the physical world. They study atoms, particles, neutrons, and electrons. They observe the ways in which moving bodies behave and stationary bodies react to pressure and other forces. They observe waves and small particles. They observe how physical forces affect living things. In short, the physical world provides a virtually unlimited store of data for the various types of physicists to observe.

Inferences, Generalizations, or Hypotheses are made regarding the scope of the phenomena. When possible, physicists seek general hypotheses or physical theories that they can test, modify, and perfect by extended study and experimentation. When successful, they predict new physical phenomena in line with a given theory and then conduct further observations or experiments to confirm or falsify it.

Implications. The huge growth in knowledge and understanding of the physical world as a result of advances in physics carries with it important implications for quality of life in many dimensions of human existence. It has provided the foundations of engineering. It enables us to build power plants, trucks, airplanes, trains, televisions, and telephones. Most machinery and tools, for example, are dependent on knowledge of physics. Most construction of buildings, irrigation and sewer systems, solar power alternatives, and the instrumentation of modern medicine are products of modern physics. Our knowledge of physics has also (arguably) been misused in the building of weapons of mass destruction, in our polluting of the environment, and in our use of mechanisms by which to invade the privacy of citizens.

The Point of View: Physicists see the universe, as well as the physical world and everything in it, as ultimately explainable and understandable through physical theories and laws. Many physicists see the universe as open to almost unlimited exploration and discovery.

The Logic of Chemistry

The Goal of Chemistry is to study the most basic elements out of which all substances are composed and the conditions under which, and the mechanisms by which, substances are transformed into new substances. Chemists study pure substances, elements and compounds, molecules, atoms, and sub-atomic particles. They study chemical reactions, classes of chemicals, and uses for chemicals. Chemistry, like physics, conducts its study of the physical properties of chemical substances insofar as the properties of these substances can be measured, expressed in mathematical formulas (or approximations), and explained by chemical theories. Its goals may be roughly contrasted with those of physics (which focuses on physical properties, on the physical nature of matter and energy).

Its Key Question is: What are the chemical properties of pure substances insofar as they can be measured, expressed in mathematical formulas, and explained by chemical theories?

Its Key Concepts: Chemical theory is based on a conception of atoms, their electronic structures, and their spatial arrangements in molecules. Other key concepts include matter, energy, gravity, physical property, chemical property, pure substance, element, compound, molecule, reaction, electron, electron transfer, electron sharing, chemical bonding, atomic weight, molecular weight, specific gravity, valence, catalysis, qualitative analysis, quantitative analysis, organic compound, and inorganic compound.

Its Key Assumptions are: That the universe is controlled by laws, that the same laws apply throughout the universe, that the laws guiding the universe can be expressed in mathematical terms and formulas, that physical properties can be distinguished from chemical ones, that all (or most) of the changes in identity of substances, as they react with other substances, can be accounted for by the theories and laws of modern chemistry.

The Data Chemists Gather result from their observations of the physical and chemical properties of matter. They observe matter as divided into elements and compounds. They seek to gather information about pure substances, molecules, atoms, and subatomic particles. They compare the behavior of different molecules. They observe the speed of chemical reactions within plants and animals. They observe the extent to which helping agents are necessary for these reactions to take place.

Inferences, Generalizations, or Hypotheses are made regarding the scope of chemical phenomena. When possible, chemists seek general hypotheses or chemical theories that they can test, modify, and perfect through extended study and experimentation. When successful, they predict new chemical phenomena in line with a given theory and then conduct further experiments to verify or falsify it.

Implications. The huge growth in knowledge and understanding of the chemical world as a result of advances in chemistry carries with it important implications for quality of life in many dimensions of human existence. Chemical knowledge has had significant implications in medicine, agriculture, engineering, and biology. Many new substances and materials have been produced through chemistry. Our knowledge of chemistry has also been misused in the building of weapons of mass destruction (biochemical weapons), in our polluting of the environment, and in creating chemicals harmful to people, other animals, and plants.

The Point of View. Chemists see the physical world as containing basic elements whose structures can be studied and transformed in accordance with various chemical laws and principles.

The Logic of Geology

Goals of Geologists: The purpose of geology is to understand the earth and all its aspects—its origin, its varied features, the composition and structure of the material that composes it, and its impact on the life upon it. It is concerned with all the forces that have acted upon the earth and the effects of those forces. It attempts to reconstruct the history of the earth, particularly as it is recorded in the rocks of the outer crust.

Questions Geologists Ask: What is the earth made of? How has the earth changed over time? What causes the earth to change? How can we predict changes in the earth? How can we use what we know about the physical environment in making decisions?

Information Geologists Use: Geologists primarily study rocks and derivative materials that make up the earth's crust, as well as information about physical forces that affect the earth's development. For example, they use information about the earth's water in relation to geological processes. They use maps of the earth, as well as knowledge from geodesy (the branch of applied math concerned with measuring, or determining the shape of the earth, or with locating points on the earth). Geologists gather information about landforms and other surface features of the earth, as well as information about minerals within the earth. Geologists study the geomagma field, paleomagnetism in rocks and soils, heat flow phenomena within the earth, chemicals within the earth, sediments, oil, coal, fossils, and geothermal energy.

Judgments Geologists Make: Geologists make judgments about the physical properties of the earth and its internal composition. They make judgments about the causes of change in the earth, the chemical makeup of the earth, the origin, structure, history and composition of rocks and minerals. They make judgments about prehistoric life, about how the earth is altered due to external forces, about how best to utilize and exploit the earth's natural resources, about how to design human-made structures given the earth's processes and makeup, and about problems caused by human use and exploitation of the physical environment.

Concepts that Guide Geologists' Thinking: Key geological concepts include: the geological time scale (obtained from four major rock types, each produced by a different kind of crustal activity; endogenetic processes (processes originating within the earth) exogenetic processes (those that originate externally); and the plate tectonics hypothesis; (that the earth's crust is divided into a number of plates that move about, collide, and separate over geological time). In addition, geologists use principles from other sciences to understand the earth. For example, biology is needed to explain life records of the past; the remote history of the earth's beginnings are interwoven with astronomy; theories in physics must be used to explain tides, earth heat, interior rigidity and many other phenomena; chemistry is needed to analyze the materials within the earth; theories in meteorology and climatology are needed to explain how external forces impact the earth's surface.

Key Assumptions Geologists Make: That geology is interwoven with many other branches of science and therefore, that geologists must rely on theories and laws from other scientific branches to think geologically; that the history of the earth is best interpreted in terms of what is known about geological processes at work in the present, rather than supposed processes in the past (principle of uniformitarianism); that the structures and forces within the earth, as well as those affecting the earth, are interrelated and must be understood in relationship to one another; that the physical structures of the earth and the ways in which they function are predictable, though there is much about the earth that we cannot yet predict.

Implications of Geology: There is almost unlimited practical value in applying geological knowledge. By studying geology, for example, there are implications for determining and predicting water, coal, and oil supply, stone quarrying, and locating ore. Geology aids in identifying geologically stable environments for human constructions. It can also help in forecasting natural hazards associated with geodynamic forces including volcanoes and earthquakes. Geologists can make a significant contribution in illuminating the effects of human exploitation on the earth's surface and resources.

The Point of View of Geologists: Geologists see the earth as a physical structure containing predictable structures, influences and forces which, when systematically studied, can improve the quality of life.

The Logic of Astronomy

Goals of Astronomers: Astronomers study the universe in order to better understand what it is comprised of and how celestial bodies and energy function within it. Astronomers seek to understand the origins, evolution, composition, motions, relative positions, size and movements of celestial bodies, including planets and their satellites, comets and meteors, stars and interstellar matter, galaxies and clusters of galaxies, black holes and magnetic fields, etc.

Questions Astronomers Ask: How did matter and energy in the universe ever come to be? How is the universe structured? What energy forces exist in the universe and how do they function? Will the universe continue to expand forever? How are celestial bodies born? How do they function? How do they evolve? How do they die? Do planets similar to earth exist in the universe? What questions remain to be asked about the universe?

Information Astronomers Use: Astronomers gather information about celestial bodies and energy through direct observation and indirect measurements. Developing methods for gathering information about the universe is a key ongoing focus of astronomers' work. For example, they use telescopes, as well as images taken from balloons and satellites. They gather information about radiation of bodies in the universe through the electromagnetic spectrum, including radio waves, ultraviolet and infra-red radiation, X-rays and gamma rays. Telescopes placed on orbiting satellites gather information about radiation blocked by the atmosphere. Astronomers rely on computers with image processing software that notes the power and shape of light. They also use the interferometer, a series of telescopes that collectively have tremendous power.

Judgments Astronomers Make: Astronomers make judgments about the universe and how it functions. Using the instruments they design and continually seek to refine, they make judgments about suns, stars, satellites, moons, nebulae and galaxies, black holes, magnetic fields, gas clouds, comets, etc. They make judgments about the distances, brightness, and composition of celestial bodies and their temperature, radiation, size, and color. From a practical perspective, astronomers make judgments that include making astronomical tables for air and sea navigation, and determining the correct time.

Concepts that Guide Astronomer's Thinking: The universe is the most fundamental concept in astronomy. The universe is the total of all bodies and energy in the cosmos that function as a harmonious and orderly system. Other important concepts in astronomy include: gravity, electromagnetism, nuclear forces (strong and weak), and quantum theory.

Key Assumptions Astronomers Make: 1) There are laws governing the universe, though we don't yet know them all; 2) The universe is largely unexplored and at present unexplained; 3) We need to develop better instruments of observation and measurement to understand the universe; and 4) Judgments in astronomy are limited by the observational instruments and research methods currently available.

Implications of Astronomy: One important implication of astronomy is that, as we improve our understanding of the universe, based on scientific observations and conclusions, we improve our understanding of life as an organic process, and we therefore rely less on myth to explain the universe. Furthermore, advances in astronomy help us see the earth as a miniscule body within a vast, expanding universe, rather than the earth (and therefore humans) as the center of the universe.

The Point of View of Astronomers: Astronomers look at the universe, and see a vast system of systems and a hugely unexplored space waiting to be discovered and understood.

The Logic of Biology

Biological Goals: Biology is the scientific study of all life forms. Its basic goal is to understand how life forms work, including the fundamental processes and ingredients of all life forms (i.e., 10,000,000 species in fragile ecosystems).

Biological Questions: The questions biology is concerned with are: What is life? How do living systems work? What are the structural and functional components of life forms? What are the similarities and differences among life forms at different levels (molecule, organelle, cell, tissue, organ, organism, population, ecological community, biosphere)? How can we understand the biological unity of living matter?

Biological Information: The kinds of information biologists seek are: information about the basic units out of which life is constructed, about the processes by which living systems sustain themselves, about the variety of living systems, and about their structural and functional components.

Biological Judgments: Biochemists seek to make judgments about the complex processes of maintenance and growth of which life basically consists.

Biological Concepts: There are a number of essential concepts to understand to understand the logic of biology: the concept of levels of organization of life processes (at the molecular level, at the sub-cellular particle level, at the cellular level, at the organ level, and at the level of the total organism), the concept of life structures and life processes, the concept of the dynamics of life, the concept of the unity of life processes amid a diversity of life forms, etc...

Biological Assumptions: Some of the key assumptions behind biological thinking are: that there are foundations to life, that these foundations can be identified, studied, described, and explained; that it is possible to use biological concepts to explain life; that it is possible to analyze and discover the structure and dynamics of living systems and their components; that all forms of life reproduce, grow, and respond to changes in the environment; that there is an intricate and often fragile relationship between all living things; that all life forms, no matter how diverse, have common characteristics: 1) they are made up of cells, enclosed by a membrane that maintains internal conditions different from their surroundings, 2) they contain DNA or RNA as the material that carries their master plan, and 3) they carry out a process, called metabolism, which involves the conversion of different forms of energy by means of which they sustain themselves.

Biological Implications: There are specific and general implications of the present logic of biology. The specific implications have to do with the kind of questioning, the kind of information-gathering and information-interpreting processes being used by biologists today. For example, the state of the field implies the importance of focusing questions and analysis on the concepts above, of seeking key answers at all levels of life systems. The general implications are that we have the knowledge, if not always the will, to understand, maintain, and protect forms of life.

Biological Point of View: The biological viewpoint is focused on all levels and forms of life. It sees all life forms as consisting in structures and understood through describable functions. It sees life processes at the molecular level to be highly unified and consistent. It sees life process at the whole-animal level to be highly diversified.

The Logic of Zoology

Goals of Zoologists: To analyze, classify, and come to understand all forms of animal life, using methods of comparative anatomy and other systemic procedures. Zoologists seek to understand the structure of animal bodies, their habits, how they live, grow and reproduce, and how they interact with plants and other animals.

Questions Zoologists Ask: What is the best system we can construct for analyzing, classifying, and understanding all forms of animal life? What forms and structures unify and differentiate animal species? What can we learn about animals from animal embryology, physiology, anatomy, ecology, and biochemistry? What is the life cycle, distribution, and evolutionary history of this or that particular animal species?

Information Zoologists Use: Is taken from field and laboratory observations of animals, particularly as they function in groups. Zoologists gather information about the structure of animals as well as processes common within animal groups. Zoologists also use basic information from biology (e.g., from cell biology, anatomy, physiology, embryology, genetics, sociobiology, and biochemistry).

Concepts that Guide Zoologists' Thinking: Include that of unifying traits and diversifying variations of animal life, including protozoa, cell, food gatherer, phylum, class, order, family, genus, species, morphology (systems of muscles or bones), histology (body tissues), cytology (cells and their components), neurology, embryological origin, animal physiology, anatomy, homeostasis, heredity, genetics, ecology, biochemistry, invertebrate zoology, entomology (insects), malacology (mollusks), vertebrate zoology, ichthyology (fish), herpetology (amphibians and reptiles), ornithology (birds), mammalogy (mammals), vertebrate zoology.

Key Assumptions Zoologists Make: That there are ways to analyze and classify all forms of animal life so as to shed light on the forms and structures that unify and differentiate them, and that by understanding the similarities and differences between them we can reconstruct evolutionary history.

Judgments Zoologists Make: Zoologists make judgments about ways in which animals are alike and different, about what makes them living things, about the foods they gather, the ways they reproduce, the ways they have evolved and are evolving, about the parasites that live within and on them, about their diseases and their health, in short, about the most significant distinctions between and among them.

Implications of Zoology: Zoology carries with it important implications for understanding the basis of evolution and the world ecosystem. Zoology has made a significant contribution to our understanding of ecosystems and their fragility, and provided insights into more sensible ways to manage agriculture, forests, and marine life. Finally, it has led to many other applications in diverse areas, including medicine.

The Point of View of Zoology: Zoologists see animals as functioning in groups in interrelationships with one another and the natural world. And they see the study of animals as vital to understanding how animals function, adapt, evolve, and survive.

The Logic of Botany

Goals of Botanists: To understand plants and all their aspects, including their life processes, structures, and growth patterns; to understand the relationships between plants, as well as between plants and animals.

Questions Botanists Ask: How do different plants function? How are they structured? How do plants differ in their needs? How do plants interact with other plants? How do they interact with the environment? How do they interact with animals? What threatens plants? How important is the native environment to the growth and propagation of plants?

Information Botanists Use: The main information used by botanists is plants themselves. They observe plants in their natural habitats. They observe them in artificial environments. They compare plants. They use information about seeds. They look at how plant cells function. They observe how plants grow. They observe how animals interact with plants. They look for pathology in plants.

Judgments Botanists Make: Botanists make judgments about the differences between plants, about how they best function and thrive, about pathologies within plants, about how they interact with other plants, about how they interact with animals.

Concepts that Guide Botanists' Thinking: The most fundamental concept in botany is the concept of photosynthesis, since plants are defined as multicellular organisms that carry out photosynthesis. Photosynthesis is the food-making process within plants, which uses sunlight as its energy source. Photosynthesis is made possible through chlorophyll, a green pigment in plants, which enables plants to make their own food using carbon dioxide, minerals, and water. Thus, chlorophyll is another key concept. Other important concepts used to guide botanical thinking are: plant anatomy (the study of internal plant structure), plant cytology (the study of plant cells), plant morphology (the study of the forms and shapes of plants), plant physiology (the study of how plants grow, breathe, make food, etc.), plant pathology (the study of plant disease), plant taxonomy (the naming of plants and plant groups), and plant ecology (the study of the plant in relationship to its habitat).

Key Assumptions Botanists Make: Botanists make the following assumptions: 1) plants should be grouped according to structure and growth patterns; 2) the land plants that we study today evolved from ancient water plants; 3) inheritance laws control the way plant parents pass certain characteristics on to their offspring; 4) the study of plants is important, regardless of any implications for human decision-making; 5) the study of plants is nevertheless necessary since all animals depend upon plants for food, and important implications follow from studying plants.

Implications of Botany: Though botany is a pure science, there are important implications of botanical thinking. Botany can make a significant difference in the preservation and enhancement of plant life. It can lead to a greater understanding of the medicinal qualities of plants, and hence to improved medicine. But it can also be inadvertently used in negative ways, as in the use of pesticides that cause damage to human and animal life.

The Point of View of Botanists: Botanists look at plants as essential to the survival of all living creatures.

The Logic of Biochemistry

Biochemical Goals: The goal of biochemistry is to determine the biological foundations of life through chemistry. Its aim is to use chemistry to study events on the scale of structures so small they are invisible even with a microscope.

Biochemical Questions: How do small-scale structures and events underlie the larger-scale phenomena of life? What chemical processes underlie living things? What is their structure? And what do they do? How can we correlate observations made at different levels of the organization of life (from the smallest to the largest)? How can we produce drugs that target undesirable events in living creatures?

Biochemical Information: The kinds of information biochemists seek are: information about the kind of chemical units out of which life is constructed, about the process by which key chemical reactions essential to the construction of life take place.

Biochemical Judgments: Biochemists seek to make judgments about the complex process of maintenance and growth of which life basically consists. In short, they seek to tell us how life functions at the chemical level.

Biochemical Ideas: There are a number of ideas essential to understanding biochemistry: the idea of levels of organization of life processes (molecular, sub-cellular particle, cellular, organ, and total organism), the idea of life structures and life processes, the idea of the dynamics of life, the idea of the unity of life processes amid a diversity of life forms, etc.

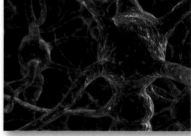

Biochemical Assumptions: Some of the key assumptions behind biochemical thinking are: that there are chemical foundations to life, that the techniques of chemistry are most fitting for the study of life at the level of molecules, that it is possible to use chemical ideas to explain life, that it is possible to analyze and discover the key agents in fundamental life process, and that it is possible, ultimately, to eliminate unwanted life processes while strengthening or maintaining desirable ones.

Biochemical Implications: The general implications of biochemistry are that we will increasingly be able to enhance human and other forms of life, and to diminish disease and other undesirable states, by application of chemical strategies.

Biochemical Point of View: The biochemical viewpoint sees the chemical level as revealing fundamental disclosures about the nature, function, and foundations of life. It sees chemistry as solving the most basic biological problems. It sees life processes at the chemical level to be highly unified and consistent, despite the fact that life process at the whole-animal level are highly diversified.

The Logic of Paleontology

The Goal of Paleontology: Is to discover the nature and implications of fossils of animal and plant life that existed in remote geological times (from 600 million years ago to 2 million years ago).

Its Key Questions are: What can we learn about the development of plant and animal life by studying the fossil remains of animal and plant life from 600 million years ago to 2 million years ago? What is the life cycle, distribution, and evolutionary history of this or that particular plant or animal species?

Its Key Concepts include: ancient life form, paleozoic life forms, plant, animal, fossil, petrification, organic material, inorganic material, natural mold, carbon print, sedimentary deposit, geological deposit, fossil animal droppings, Cambrian period (600 million years ago), vertebrates, primitive forms of crustacean, mollusks, Ordovician period (500 million years ago), graptolites, colonial coelenterates, primitive fish, flora, fauna, Silurian period (430 million years ago), fish, scorpion, vacular plants, corals, Devonian period (395 million years ago) amphibians, Carboniferous period (345 million years ago) lizards and sharks, Permian period (280 million years ago) early reptiles, Mesozoic life, age of reptiles (begins 225 million years ago) Triassic period (225 million years ago) dinosaurs, Jurassic period (195 million years ago) dinosaurs, Cretaceous period (136 million years ago) horned dinosaurs, Cenozoic life, age of mammals (begins 65 million years ago), Eocene epoch (54 million years ago), Oligocene epoch (38 million years ago), Miocene epoch (26 million years ago), and Pleistocene epoch (12 million years ago).

Its Key Assumptions are: That plant and life forms originated on earth, that this evolution took place over some 600 million years; that paleontology can be understood through studying fossil remains from distinctive periods of time in that 600 million years; that paleontology gives a true but incomplete record of the development of existing life forms; and that throughout geological time successive plants and animals have tended to become more complex.

The Data or Information Paleontologists Gather are of and from the actual remains of living organisms preserved and protected from decay by enclosure in the crust of the earth through fossilization (ancient plants and animals embedded in mineral deposits). Paleontologists rely on basic information from geology and biology in conducting their investigations.

Inferences, Generalizations, or Hypotheses are made regarding the ancient evolution of plant and life forms. It is in general inferred that life in the sea evolved from primitive multicellular free-floating forms to advanced groups capable of life on land (from fossil remains in rock strata of the Paleozoic era) and that each group of animals begins with simple types, that gradually more complex and specialized forms appear, and that frequently decadence sets in with great suddenness, resulting in extinction or the reduction of the group to relative unimportance.

Implications: The huge growth in knowledge and understanding of the evolution of the plant and animal world as a result of advances in Paleontology carries with it important implications for understanding the basis for human evolution. Paleontology has also made a significant contribution to our understanding of ecosystems and their fragility. Certain fossils are so characteristic of the different periods, epochs, or formations of rocks that they serve as index fossils enabling the geologist to fix the geological age of the rocks from which they come.

The Point of View of Paleontology: Paleontologists see the development of plants and animals occurring over millions of years and the study of this evolutionary process as an ongoing, dynamic process.

The Logic of Animal Physiology

Goals of Animal Physiologists: To figure out how living organisms work, including the physical and chemical processes that take place in living organisms during the performance of life functions. Physiology investigates biological mechanisms with the tools of physics and chemistry. It is closely related to anatomy, though physiologists focus on bodily functions; anatomists on bodily structures. General physiologists focus on the basic functions common to all life. Physiologists may focus on particular life forms, pathology, or comparative studies. (Plant physiology, a branch of botany, focuses on the life functions within plants.)

Questions Animal Physiologists Ask: What are the basic functions common to all life? What physical and chemical processes take place in living organisms during the performance of life functions? What happens in a body during reproduction, growth, metabolism, respiration, digestion, excitation, and contraction? What happens during these functions within the bodies' cells, tissues, organs, or within organ or nerve systems? In what ways can life functions be disrupted, injured, or diseased?

Information Animal Physiologists Use: The main information obtained by physiologists is from the direct study of physical and chemical processes as these take place in living organisms during the performance of life functions. They observe cells, tissue, and organs microscopically. They observe them in artificial and real-life environments. They compare structures and functions of life processes.

Judgments Animal Physiologists Make: Physiologists make judgments about functions common to all life forms as well as differences among them. They judge how these functions best perform and thrive. They make judgments about pathologies and interpret how internal systems and functions interrelate with environment realities.

Concepts that Guide Animal Physiologists' Thinking: The most fundamental concept in physiology is the concept of bodily functions in systemic relations. Other important concepts used to guide physiological thinking are: reproduction, growth, metabolism, respiration, blood circulation, nutrition, digestion, excretion, excitation, contraction, cells, tissues, nerves, muscles, bones, organs, systems of organs, organ and system pathology.

Key Assumptions Animal Physiologists Make: Physiologists make the following assumptions: 1) living things must perform a specifiable group of common and essential functions; 2) different species of living things perform various common functions in different and sometimes unique ways (through diverse cell, tissue, and organ structures); and, 3) it is possible for physiologists to accurately describe, test, and verify their descriptions and theories concerning functions performed within animal systems.

Implications of Animal Physiology: The implications of human physiology are interconnected with those of bacteriology, immunology, chemistry, and physics, among other scientific branches. Physiologists who study animal functions have made numerous discoveries about bodily functions (such as the heart, brain,and other organs) which have led to advancements in medical treatment. The implications for medical care, for human and veterinary medicine, through physiological study, are virtually unlimited.

The Point of View of Physiologists: Physiologists see life functions as systems working harmoniously to perform essential processes. They also see pathology within living systems as a breakdown in this harmonious process which, when studied, can lead to less pathology and improved life quality.

The Logic of Archaeology

Purpose of the Thinking: The purpose of archaeology is to find remnants of the past, interpreting and piecing them together in order to discover more about historical events, culture, and our human legacy.

Question at Issue: What is the best way to find information about the distant past, and how does one effectively interpret the past through archaeology?

Information: In order to become or think like an effective archaeologist, one should consider site discovery techniques, artifact retrieval, cataloging, and preservation techniques, contextual and cultural clues, and supportable historical and scientific data from archaeological finds.

Interpretation and Inference: Archaeologists formulate historical interpretations and validate them by cross-referencing various previous interpretations, current cultural evidence, physical artifacts and scientific data from archaeological finds.

Concepts: The concept of recovering lost history, of seeking evidence from beneath the surface of the earth to reveal important events and time sequences in ancient human history.

Assumptions: We can always enrich our understanding of the past, and archaeology provides evidence to support historical theories. The past is a puzzle that can be further solved through ongoing archaeological study.

Implications and Consequences: New discoveries that answer questions of the past can be made with on-going archaeological research. Beliefs we now hold as true, could one day be revised based on future discoveries. Understanding old ways of doing things may also provide the present or future with useful knowledge or resources for survival.

Point of View: Seeing the story of humankind as taking place through stages over hundreds of thousands of years.

The Logic of Ecology

Goals of Ecologists: Ecologists seek to understand plants and animals as they exist in nature, with emphasis on their interrelationships, interdependence, and interactions with the environment. They work to understand all the influences that combine to produce and modify an animal or given plant, and thus to account for its existence and peculiarities within its habitat.

Questions Ecologists Ask: How do plants and animals interact? How do animals interact with each other? How do plants and animals depend on one another? How do the varying ecosystems function within themselves? How do they interact with other ecosystems? How are plants and animals affected by environmental influences? How do animals and plants grow, develop, die, and replace themselves? How do plants and animals create balances between each other? What happens when plants and animals become unbalanced?

Information Ecologists Use: The primary information used by ecologists is gained through observing plants and animals themselves, their interactions, and how they live within their environments. Ecologists note how animals and plants are born, how they reproduce, how they die, how they evolve, and how they are affected by environmental changes. They also use information from other disciplines including chemistry, meteorology, and geology.

Judgments Ecologists Make: Ecologists make judgments about how ecosystems naturally function, about how animals and plants within them function, about why they function as they do. They make judgments about how ecosystems become out of balance and what can be done to bring them back into balance. They make judgments about how natural communities should be grouped and classified.

Concepts that Guide Ecologists' Thinking: One of the most fundamental concepts in ecology is ecosystem. Ecosystem is defined as a group of living things, dependent on one another and living in a particular habitat. Ecologists study how differing ecosystems function. Another key concept in ecology is ecological succession, the natural pattern of change occurring within every ecosystem when natural processes are undisturbed. This pattern includes the birth, development, death, and then replacement of natural communities. Ecologists have grouped communities into larger units called biomes. These are regions throughout the world classified according to physical features, including temperature, rainfall, and type of vegetation. Another fundamental concept in ecology is balance of nature, the natural process of birth, reproduction, eating and being eaten, which keeps animal/plant communities fairly stable. Other key concepts include imbalances, energy, nutrients, population growth, diversity, habitat, competition, predation, parasitism, adaptation, coevolution, succession and climax communities, and conservation.

Key Assumptions Ecologists Make: That patterns exist within animal/plant communities; that these communities should be studied and classified; that animals and plants often depend on one another and modify one another; and that balances must be maintained within ecosystems.

Implications of Ecology: The study of ecology leads to numerous implications for life on earth. By studying balance of nature, for example, we can see when nature is out of balance, as in the current population explosion. We can see how pesticides, designed to kill pests on farm crops, also lead to the harm of mammals and birds, either directly or indirectly through food webs. We can also learn how over-farming causes erosion and depletion of soil nutrients.

The Point of View of Ecologists: Ecologists look at plants and animals and see them functioning in relationship with one another within their habitats, and needing to be in balance for the earth to be healthy and sustainable.

The Problem of Pseudo-Scientific and Unscientific Thinking

Unscientific and pseudo-scientific thinking come from the unfortunate fact that humans do not naturally think scientifically, though they often think they do. Furthermore, we become explicitly aware of our unscientific thinking only if trained to do so. We do not naturally recognize our assumptions, the unscientific way we use information, the way we interpret data, the source of our concepts and ideas, the implications of our unscientific thought. We do not naturally recognize our unscientific perspective.

As humans we live with the unrealistic but confident sense that we have fundamentally figured out the true nature of things, and that we have done this objectively. We naturally believe in our intuitive perceptions — however inaccurate. Instead of using intellectual standards in thinking, we often use self-centered psychological standards to determine what to believe and what to reject. Here are the most commonly used psychological standards in unscientific human thinking.

"IT'S TRUE BECAUSE I BELIEVE IT." I assume that what I believe is true even though I have never questioned the basis for my beliefs.

"IT'S TRUE BECAUSE WE BELIEVE IT." I assume that the dominant beliefs within the groups to which I belong are true even though I have never questioned the basis for many of these beliefs.

"IT'S TRUE BECAUSE I WANT TO BELIEVE IT." I believe in, for example, accounts of behavior that put me (or the groups to which I belong) in a positive rather than a negative light even though I have not seriously considered the evidence for the more negative account. I believe what "feels good," what supports my other beliefs, what does not require me to change my thinking in any significant way, what does not require me to admit I have been wrong.

"IT'S TRUE BECAUSE I HAVE ALWAYS BELIEVED IT." I have a strong desire to maintain beliefs that I have long held, even though I have not seriously considered the extent to which those beliefs are justified, given the evidence.

"IT'S TRUE BECAUSE IT IS IN MY INTEREST TO BELIEVE IT." I hold fast to beliefs that justify my getting more power, money, or personal advantage even though these beliefs are not grounded in sound reasoning or evidence.

Since humans are naturally prone to assess thinking in keeping with the above criteria, it is not surprising that unscientific thinking flourishes in our society.

A Pseudo-Science:
Why Astrology Is Not a Science[2]

The claims of astrologers are rejected by the scientific community. Astrology is popular, nevertheless. Though most adults have taken many science classes in school, few know how to assess the claims of astrologers scientifically. In fact, many students, and even teachers, believe that astrology provides accurate personality descriptions and valuable advice. Noted astrologers earn a sizeable income as consultants. To many, the personality descriptions based on horoscopes seem to fit. As people read the descriptions of personality traits attributed to those born under their "sun sign," they examine themselves and find they have many of the traits depicted. What they do not do is look to see if descriptions associated with other signs of the Zodiac might fit them equally well. Likewise, when they are told that at the present time in their lives they are going through some stress or have to make a major decision, they tend to agree with the astrologer, without examining their lives to see if the same description would fit almost any other period of their lives.

Simply telling students that most scientists consider astrology invalid will not convince them that it is. After learning about controlled research, students should be able to see the defects in conventional astrological research. They should also be able to identify research designs capable of scientifically testing astrological theories.

Scientists agree that the positions of the sun, moon, and possibly even some nearby planets affect living organisms—but not in the ways claimed by astrologers. Carefully controlled studies of predictions based on astrological theories have almost always yielded negative results.

Astrology began thousands of years ago in ancient Babylonia, Persia, Greece, and Rome. Before true scientific knowledge existed, and before what we call the 'scientific method' was devised, these people tried to organize their knowledge of the stars by perceiving in them shapes of animals and persons, such as a lion, ram, crab, fish, scorpion, archer, water carrier, twins, etc. The ancients assumed that the arrangement of stars into the shapes of animals and persons had cosmic significance.

They noticed that during the day the sun passed through the areas in which these shapes were observed at night, and this varied at different times of the year. The band of these shapes that the sun passed through was called the 'Zodiac,' and these animals or persons were called the 'signs of the Zodiac.' For awhile the sun was in the area of a constellation shaped like fishes, a month later the sun would be in a constellation which had stars that reminded them of a water carrier. It was believed that constellations were powerful when the sun was in their area. Thus if the sun was in the constellation shaped like a lion, this cosmic animal would have a powerful influence on earthly events.

---

2   These ideas were originally formulated by Dr. Wesley Hiler.

These ancients noticed that some lights in the sky moved across the stable arrangement of the other lights, so they called these 'wanderers' or 'planets' and imagined that they were gods or the abodes of gods. The sun was worshiped as the chief god in some of these lands. They noticed that one of the planets was reddish in color so they named it after the god of war, Mars.

Astronomers in these ancient civilizations assumed that the arrangement of stars, planets, sun, and moon influenced events taking place on earth at the time. Specifically the arrangement of these heavenly bodies at the time of an individual's birth would influence his or her personality and fate. The arrangement of heavenly bodies at any given time is called a 'horoscope.' It was assumed that if astrologers knew the time and place of an individual's birth they could make predictions concerning that person's character and destiny. For instance, if a person was born when the sun was in the part of the sky where the stars were arranged in the shape of a lion, the person would have personality traits similar to those of a lion. The region of the Zodiac where the sun was at the time of a person's birth is referred to as the individual's 'sun sign.' Sun signs are the most commonly used sources of astrological inferences. Many newspapers contain advice geared to a person's sun sign.

The theories developed by ancient astrologers were passed on through tradition, without any carefully controlled scientific verification, generation after generation. Because of the enormous number of variables in a horoscope, and the many possible ways of interpreting each one, an astrologer can select the interpretation he believes best fits a known individual. Therefore astrologers are quite accurate in matching their predictions with famous people of the past whose time and place of birth are known to them. They are less accurate in using horoscopes to make inferences about the personalities and lives of people unknown to them. Most books on astrology contain chapters on famous people like Hitler or Napoleon, in which the astrologer is able to match their lives with inferences derived from the arrangement of the stars at the times of their births and at times during their lives.

Astrological method differs from scientific method in many ways:

I) Astrological interpretations are not derived from natural laws but from symbolic relationships. According to astrology, a person born when the sun, moon, or any planets were in a constellation which looked like a ram would have personality traits similar to a ram.

2) Astrologers seek correspondences between astrological theory and what is known about someone and ignore lack of correspondence.

3)   Astrologers do not conduct carefully controlled research to see if their personality assessments and predictions are more accurate than one could expect by chance alone.

4)   Some personality descriptions are so general that they fit everyone. Everyone has some traits of all the sun signs, so people can find descriptions which fit them in every sun sign.

5)   People can find any tendency in themselves if they look hard enough. They see what they expect to see. Their knowledge of astrology affects how they see themselves.

6)   People jump to conclusions on the basis of small samples. They tend to remember what fits their expectations, but forget what doesn't.

How could the arrangement of stars as seen from the earth could have any effect on events taking place on earth; for instance, how could a lion shaped arrangement of stars influence events in a lion like fashion? Nevertheless, Sydney Omarr, a well known and highly influential astrologist, wrote books on the twelve signs of the zodiac which sold 50 million copies world-wide.[3]

3   *The Press Democrat*, Jan. 3, 2003, p. B2.

A Critical Approach to Scientific Thinking*

A critical approach to learning science is concerned less with accumulating undigested facts and scientific definitions and procedures, than with learning to *think scientifically*. As we learn to think scientifically, we inevitably organize and internalize facts, learn terminology, and use scientific procedures. But we learn them deeply, because they are tied into ideas we have thought through, and hence do not have to "re-learn" later.

The biggest obstacle to science education for most of us is our previous conceptions. Growing up we develop our own ideas about the physical world — many of which are false. What school texts say usually does not change our inner beliefs. Unscientific beliefs continue to exist in an unarticulated and therefore unchallenged form in our minds. For example, in one study, few college physics students could correctly answer the question, "What happens to a piece of paper thrown out of a moving car's window?" They reverted to a naïve version of physics inconsistent with what they learned in school; they used Aristotelian rather than Newtonian physics. *The Proceedings of the International Seminar on Misconceptions in Science and Mathematics* offers another example. A student was presented with evidence about current flow that was incompatible with his articulated beliefs. In response to the instructor's demonstration, the student replied, "Maybe that's the case here, but if you'll come home with me you'll see it's different there."[4] This student's response  graphically illustrates one way we retain their own beliefs: we simply juxtapose them with a new belief. Unless we practice expressing and defending our scientific beliefs, and listening to those of others, we will not critique our own beliefs and modify them in accordance with what we learn while studying science.

Typical science texts present the finished products of science. Students are often required to practice the skills of measuring, combining liquids, graphing, and counting, but see no reason behind it. The practice becomes mindless drill.

Texts also introduce scientific concepts. Yet few people can explain scientific concepts in ordinary language. For example, after a unit on photosynthesis, it would not be uncommon for students who were asked, "Where do plants get their food?" to reply, "From water, soil, and all over." Students often misunderstand what the concept 'food' means for plants (missing the crucial idea that *plants make their own food*).

---

* The ideas on these two pages were originally formulated by Dr. Wesley Hiler.

4 Hugh Helm & Joseph D. Novak, "A Framework for Conceptual Change with Special Reference to Misconceptions," Proceedings of the International Seminar on Misconceptions in Science and Mathematics, Cornell University, Ithaca, NY, June 20–22, 1983, p. 3.

Students need to understand the reasons for doing experiments or for doing them in a particular way. Lab texts often fail to make explicit the link between observation and conclusion. Rarely do students have occasion to ask, "How did we get from that observation to that conclusion?" The result is that scientific reasoning remains a mystery to most non-scientists.

A critical approach to learning science requires us to ponder questions, propose solutions, and think through possible experiments.

Many texts treat the concept of "*the* scientific method" in a misleading way. Not all scientists do the same kinds of things—some experiment, others don't, some do field observations, others develop theories. Compare what chemists, theoretical physicists, zoologists, and paleontologists do. Furthermore, scientific thinking is not a matter of running through a set of steps one time. Rather it is a kind of thinking in which we continually move back and forth between questions we ask about the world and observations we make, and experiments we devise to test out various hypotheses, guesses, hunches, and models. We continually think in a hypothetical fashion: "If this idea of mine is true, then what will happen under these or those conditions? Let me see, suppose we try this. What does this result tell me? Why did this happen? If *this* is why, *then* that should happen when I…" It is more important for students to get into the habit of thinking scientifically than to get the correct answer through a rote process they do not understand. The essential point is this: we should learn to do our own thinking about scientific questions from the start. Once we give up on trying to do our own scientific thinking and start passively taking in what textbooks tell us, the spirit of science, the scientific attitude and frame of mind, is lost.

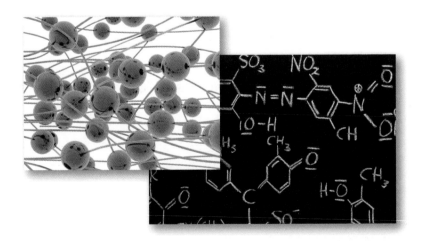

Ethics and Science

The work of science, even pure science, has implications for helping or harming living creatures, and for improving or diminishing the quality of life on earth. Scientists are increasingly concerned with the ethical implications of scientific discoveries and inventions, and with the potential of science for both good and ill.

Einstein himself underwent a transformation in his views regarding the ethical responsibilities of scientist. "From regarding scientists as a group almost aloof from the rest of the world, he began to consider them first as having responsibilities and rights level with the rest of men, and finally as a group whose exceptional position demanded the exercise of exceptional responsibilities[5]." In 1948, after the United States dropped atomic bombs on Hiroshima and Nagasaki, Einstein wrote this message to the World Congress of Intellectuals:

> We scientists, whose tragic destiny it has been to help make the methods of annihilation ever more gruesome and more effective, must consider it our solemn and transcendent duty to do all in our power in preventing these weapons from being used for the brutal purpose for which they were invented. What task could possibly be more important to us? What social aim could be closer to our hearts?[5]

Increasingly, scientists are seeking ways to use their consciences, knowledge, and unique insights to influence social and political decision-making. For example, The Union of Concerned Scientists (UCS)[6], established in 1968, works regularly to improve the ways in which science is used so as to protect the earth and it's resources. It focuses on five major areas of concern: food, vehicles, environment, energy, and security. In 1992, through its auspices, it issued a "World Scientists' Warning to Humanity," a statement supported by some 1,700 of the world's leading scientists, including most of the Nobel laureates in the sciences. The introduction reads as follows:

> Human beings and the natural world are on a collision course. Human activities inflict harsh and often irreversible damage on the environment and on critical resources. If not checked, many of our current practices put at serious risk the future that we wish for human society and the plant and animal kingdoms, and may so alter the living world that it will be unable to sustain life in the manner that we know. Fundamental changes are urgent if we are to avoid the collision our present course will bring about.

Here are some of the particular concerns of UCS:

- **The Atmosphere:** Ozone depletion, caused by pollution, results in enhanced ultraviolet radiation, which can be damaging or lethal to many life forms. It has already caused widespread damage to humans, forests, and crops.
- **Water Resources:** Heedless exploitation of ground water supplies endangers food production and other essential human systems. Some 80 countries, containing 40% of the world's population, now experience water shortages as a result of heavy demand on water resources.

---

[5] Clark, R. (1984). *Einstein: The Life and Times.* NY, NY: Avon Books, p. 723.
[6] For more information about The Union of Concerned Scientists: Citizens and Scientists for Environmental Solutions, visit Web site: (www.ucsusa. org).

- **Oceans:** Destructive pressure on the earth's oceans is severe, due to excess fishing and pollution caused by industrial, municipal, agricultural and livestock waste.
- **Soil:** Current agricultural and animal husbandry has caused extensive land abandonment, significantly reducing soil productivity.
- **Forests:** Tropical rain forests, as well as tropical and temperate dry forests, are being rapidly destroyed, resulting in the endangerment or extinction of plant and animal species.
- **Living Species:** The irreversible loss of species, which may reach 1/3 of all species now living by the year 2100, is a serious problem we now face. Our massive and heedless tampering with the world's interdependent web of life could lead to unpredictable collapses of critical biological systems whose interrelationships we are only beginning to understand
- **Population:** Pressures resulting from unrestrained population growth are resulting in an inability to sustain the earth's resources. And this problem will only worsen if current practices continue.
- **Security:** Success in reducing destruction to the earth's resources will require great reduction in violence and war. Resources now devoted to war—amounting to over $1 trillion annually—are badly needed to sustain and improve the quality of life on earth.

Critical Thinking and the Ethical Dimensions of Science

Just as we need scientists who think critically about the ethical implications of science, so we need non-scientists who think critically about the role science is playing in our world. All conscientious citizens, whether they are scientists or not, should learn to think through the ethical implications of public policies and issues—in such diverse areas as law, politics, business, medicine, biology, chemistry, engineering, and technology.

Scientific data and discovery are the result of human agency. They presuppose a social investment. They require support. If we focus scientific research in one direction, then there are other directions that receive less of a focus — less funding, less research, fewer discoveries. Science costs time, energy, and money. The science we need — if we are to eliminate starvation, disease, and environmental destruction — is not a given. We must argue for it persuasively, if that is what we want.

The implication is that we must all learn to be alert and well-informed citizens, citizens who recognize faulty reasoning, propaganda, and media bias, citizens who think independently and critically about public issues. Sound social and political thinking and enlightened scientific thinking can converge. It is in all of our interests that thist convergence takes place.

The Thinker's Guide Library

The Thinker's Guide series provides convenient, inexpensive, portable references that students and faculty can use to improve the quality of studying, learning, and teaching. Their modest cost enables instructors to require them of all students (in addition to a textbook). Their compactness enables students to keep them at hand whenever they are working in or out of class. Their succinctness serves as a continual reminder of the most basic principles of critical thinking.

For Students & Faculty

 Critical Thinking—The essence of critical thinking concepts and tools distilled into a 22-page pocket-size guide. **#520m**

 Analytic Thinking—This guide focuses on the intellectual skills that enable one to analyze anything one might think about — questions, problems, disciplines, subjects, etc. It provides the common denominator between all forms of analysis. **#595m**

Asking Essential Questions—Introduces the art of asking essential questions. It is best used in conjunction with the Miniature Guide to Critical Thinking and the Thinker's Guide on How to Study and Learn. **#580m**

 How to Study & Learn—A variety of strategies—both simple and complex—for becoming not just a better student, but also a master student. **#530m**

 How to Read a Paragraph—This guide provides theory and activities necessary for deep comprehension. Imminently practical for students. **#525m**

 How to Write a Paragraph—Focuses on the art of substantive writing. How to say something worth saying about something worth saying something about. **#535m**

 The Human Mind—Designed to give the reader insight into the basic functions of the human mind and to how knowledge of these functions (and their interrelations) can enable one to use one's intellect and emotions more effectively. **#570m**

 Foundations of Ethical Reasoning—Provides insights into the nature of ethical reasoning, why it is so often flawed, and how to avoid those flaws. It lays out the function of ethics, its main impediments, and its social counterfeits. **#585m**

 How to Detect Media Bias and Propaganda—Helps readers recognize bias and propaganda in the daily news so they can reasonably determine what media messages need to be supplemented, counter-balanced or thrown out entirely; focuses on the logic of the news and societal influences on the media. **#575m**

 Scientific Thinking—The essence of scientific thinking concepts and tools. It focuses on the intellectual skills inherent in the well-cultivated scientific thinker. **#590m**

 Fallacies: The Art of Mental Trickery and Manipulation—Introduces the concept of fallacies and details 44 foul ways to win an argument. **#533m**

For Students & Faculty, cont.

Engineering Reasoning—Contains the essence of engineering reasoning concepts and tools. For faculty it provides a shared concept and vocabulary. For students it is a thinking supplement to any textbook for any engineering course. **#573m**

Glossary of Critical Thinking Terms & Concepts—Offers a compendium of more than 170 critical thinking terms for faculty and students. **#534m**

Aspiring Thinker's Guide to Critical Thinking—Introduces critical thinking using simplified language (and colorful visuals) for students. It also contains practical instructional strategies for fostering critical thinking. **#554m**

Clinical Reasoning—Introduces the clinician or clinical student to the foundations of critical thinking (primarily focusing on the analysis and assessment of thought), and offers examples of their application to the field. **#564m**

Critical and Creative Thinking—Focuses on the interrelationship between critical and creative thinking through the essential role of both in learning. **#565m**

Intellectual Standards—Explores the criteria for assessing reasoning; illuminates the importance of meeting intellectual standards in every subject and discipline. **#593m**

Historical Guide—Focuses on history as a mode of thinking; helps students see that every historical perspective can be analyzed and assessed using the tools of critical thinking; develops historical reasoning abilities and traits. **#575m**

For Faculty

Active and Cooperative Learning—Provides 27 simple ideas for the improvement of instruction. It lays the foundation for the ideas found in the mini-guide *How to Improve Student Learning*. **#550m**

Critical Thinking Competency Standards—Provides a framework for assessing students' critical thinking abilities. **#555m**

Critical Thinking Reading and Writing Test—Assesses the ability of students to use reading and writing as tools for acquiring knowledge. Provides grading rubrics and outlines five levels of close reading and substantive writing. **#563m**

Educational Fads—Analyzes and critiques educational trends and fads from a critical thinking perspective, providing the essential idea of each one, its proper educational use, and its likely misuse. **#583m**

How to Improve Student Learning—Provides 30 practical ideas for the improvement of instruction based on critical thinking concepts and tools. **#560m**

Socratic Questioning—Focuses on the mechanics of Socratic dialogue, on the conceptual tools that critical thinking brings to Socratic dialogue, and on the importance of questioning in cultivating the disciplined mind. **#553m**